Internet Telephony For Du

Quick start on finding Internet telephony and video software on the Web

Product	Internet Download Site
Being There	http://www.beingthere.com
ClearPhone	http://www.clearphone.com
Connectix VideoPhone	http://www.connectix.com
CoolTalk	http://www.netscape.com
CU-SeeMe	http://cu-seeme.cornell.edu
DigiPhone	http://www.planeteers.com
Enhanced CU-SeeMe	http://www.cu-seeme.com
FreeTel	http://www.freetel.inter.net
FreeVue	http://www.freevue.com
...For Dummies Home Page	http://www.dummies.com
Intel Internet Video Phone	http://www.intel.com
IWave	http://www.vocaltec.com
NetMeeting Phone	http://www.microsoft.com/netmeetingInternet http://www.vocaltec.com
PGPfone	http://www.pgp.com
RealPlayer	http://www.real.com
SpeakFreely	http://www.fourmilab.ch/speakfree/windows
StreamWorks	http://www.xingtech.com
TeleChoice	http://www.telechoice.com
TeleVox	http://www.voxware.com
ToolVox	http://www.voxware.com
TrueSpeech	http://www.dspg.com
VDOLive	http://www.vdo.net
VDOPhone	http://www.vdo.net
Vidcall	http://access.digex.net/~vidcall/vidcall.html
WebPhone	http://www.netspeak.com

Cheat Sheet $2.95 value. Item 174-7.
For more information about IDG Books, call 1-800-762-2974.

...For Dummies: #1 Computer Book Series for Beginners

Internet Telephony For Dummies®, 2nd Edition

Cheat Sheet

Important Internet telephony and video terms

bandwidth: The amount of data that can flow over a connection (such as your modem connection to your Internet Service Provider) in a fixed amount of time (usually measured in kilobits or megabits per second). You can think of bandwidth as being analogous to a pipe — the bigger the pipe, the greater the volume of flow through it at once.

codec: (COmpression/DECompression algorithm) The technique used by your telephony or video program to squeeze your voice or video signal down to a small enough size so that the signal fits through the bandwidth available to your computer. Programs that use different codecs are not compatible with each other — luckily, the H.323 standard requires programs to have a common set of codecs to ensure interoperability.

directory service: A "White Pages" for the Internet. A directory service stores information (such as name, e-mail address, and IP address) about Internet telephony users in a database which other users can access. Directory services make it easier to find people to call, and help to provide current IP addresses for users whose Internet Service Providers give them a different one each time they log on.

full duplex: A full-duplex conversation enables both parties to speak simultaneously, as if speaking to a person directly or over a telephone.

gateway: A computer device that converts standard telephone and fax signals into packets for transmission over the Internet, or any TCP/IP network. A gateway can be used to connect these devices to a PC-based Internet phone, or — in conjunction with another gateway — to connect with other telephones and fax machines using the Internet instead of the telephone network to carry the signals.

half duplex: A half-duplex conversation is like talking over a walkie-talkie — only one party can speak at a time. Think of saying "over" after you finish speaking, and you've got the idea.

H.323: A standard for multimedia communications (audio, video, collaboration) over packet-based networks, especially the Internet and corporate intranets. H.323 is the standard for Internet telephony, and an increasing number of products have been designed to use the standard while communicating — allowing interoperability between different products and platforms.

host name: A plain-language name (well, almost) for your computer on the Internet. A host name (for example, `homer.simpson.edu`) corresponds to a single IP address and in most cases, you can use it in place of an IP address. Although it may look similar, your IP address differs from your e-mail address. For example, your e-mail address may be `homer@simpson.edu` and your IP address `homer.simpson.edu`, but these are not interchangeable, so don't confuse them.

intranet: In general terms, a private (or enterprise) data network, within an organization, which utilizes standard Internet protocols and systems. An intranet is typically connected to the Internet through a firewall system, which restricts the flow of data in and out of the intranet for security purposes. Internet telephony software and hardware is often used within an intranet, although communicating with the outside world usually requires special configuration of the firewall.

IP address: A four-part number that uniquely identifies your computer on the Internet (for example, 123.123.123.123). Many people who have dial-up connections to the Internet have dynamically assigned IP addresses that change every time they log on.

RTP (Real-Time Transport Protocol): Actually a pair of protocols (RTP and RTCP, or Real-Time Transport Control Protocol), RTP is a standard protocol (part of H.323) that governs how real-time multimedia data is transmitted over the Internet, and adds additional functions (such as timing control) to TCP and UDP to increase the quality of audio and video sent over the Internet.

TCP (Transmission Control Protocol): A common protocol for controlling the flow of data across the Internet. TCP provides error correction (that is, it determines if data has been lost, and then has any missing information resent), which means it often takes more time to send data from one point to another. For this reason, many Internet telephony and video programs do not use TCP.

UDP (User Datagram Protocol): Another common protocol for transmission of data across the Internet. Unlike TCP, UDP does not perform error correction, so it provides faster, although less accurate, transmission. Many Internet telephony and video programs use UDP in the interest of making transmission times faster.

INTERNET TELEPHONY FOR DUMMIES®

2ND EDITION

by Daniel D. Briere,
Patrick J. Hurley, and Rebecca Wetzel

Foreword by Tom Evslin
Vice President, AT&T WorldNet[SM] Service

IDG Books Worldwide, Inc.
An International Data Group Company

Foster City, CA ♦ Chicago, IL ♦ Indianapolis, IN ♦ Southlake, TX

Internet Telephony For Dummies®, 2nd Edition

Published by
IDG Books Worldwide, Inc.
An International Data Group Company
919 E. Hillsdale Blvd.
Suite 400
Foster City, CA 94404
www.idgbooks.com (IDG Books Worldwide Web site)
www.dummies.com (Dummies Press Web site)

Library of Congress Catalog Card No.: 97-73293

ISBN: 0-7645-0174-7

Printed in the United States of America

10 9 8 7 6 5 4 3 2 1

1DD/RX/QX/ZX/IN

Distributed in the United States by IDG Books Worldwide, Inc.

Distributed by Macmillan Canada for Canada; by Transworld Publishers Limited in the United Kingdom; by IDG Norge Books for Norway; by IDG Sweden Books for Sweden; by Woodslane Pty. Ltd. for Australia; by Woodslane Enterprises Ltd. for New Zealand; by Longman Singapore Publishers Ltd. for Singapore, Malaysia, Thailand, and Indonesia; by Simron Pty. Ltd. for South Africa; by Toppan Company Ltd. for Japan; by Distribuidora Cuspide for Argentina; by Livraria Cultura for Brazil; by Ediciencia S.A. for Ecuador; by Addison-Wesley Publishing Company for Korea; by Ediciones ZETA S.C.R. Ltda. for Peru; by WS Computer Publishing Corporation, Inc., for the Philippines; by Unalis Corporation for Taiwan; by Contemporanea de Ediciones for Venezuela; by Computer Book & Magazine Store for Puerto Rico; by Express Computer Distributors for the Caribbean and West Indies. Authorized Sales Agent: Anthony Rudkin Associates for the Middle East and North Africa.

For general information on IDG Books Worldwide's books in the U.S., please call our Consumer Customer Service department at 800-762-2974. For reseller information, including discounts and premium sales, please call our Reseller Customer Service department at 800-434-3422.

For information on where to purchase IDG Books Worldwide's books outside the U.S., please contact our International Sales department at 415-655-3200 or fax 415-655-3295.

For information on foreign language translations, please contact our Foreign & Subsidiary Rights department at 415-655-3021 or fax 415-655-3281.

For sales inquiries and special prices for bulk quantities, please contact our Sales department at 415-655-3200 or write to the address above.

For information on using IDG Books Worldwide's books in the classroom or for ordering examination copies, please contact our Educational Sales department at 800-434-2086 or fax 817-251-8174.

For press review copies, author interviews, or other publicity information, please contact our Public Relations department at 415-655-3000 or fax 415-655-3299.

For authorization to photocopy items for corporate, personal, or educational use, please contact Copyright Clearance Center, 222 Rosewood Drive, Danvers, MA 01923, or fax 508-750-4470.

About the Authors

Danny Briere lives on an island near Portland, Maine, and is President of TeleChoice, Inc., a telecommunications consulting and market research company that he founded in 1985. Author of several books on telecommunications, he is also a contributing editor for *Network World* and *Tele.Com* and has written hundreds of articles and columns for almost every telecommunications industry publication. Additionally, he is the on-air telecommunications analyst for CNNfn.

Danny and the staff at TeleChoice design new products and services for major local, long-distance, Internet, cable, and other telecommunications companies.

Danny is a graduate of the Fuqua School of Business at Duke University and also went to Duke as an undergraduate, majoring in Telecommunications Policy.

Pat Hurley lives in Coronado, California, and is a consultant and author specializing in Internet communications issues. He has written or contributed to many articles for major telecommunications industry publications, though he believes he types too slowly to ever catch up to Danny.

Pat spends much of his time working with TeleChoice, breaking in new products and services designed for Internet applications.

Pat is a graduate of Duke University, where he received a degree in Telecommunications Policy and Economics.

Rebecca Wetzel lives in Upton, Massachusetts, and is Director of Internet Consulting at TeleChoice, Inc. She has been developing and marketing Internet-based services since long before the Internet became a household word. She has written countless reports on market trends in internetworking, and is a contributor to *Network World* and *Inter@ctive Week*.

Rebecca provides marketing expertise and advice to prospective and existing Internet service and product providers and other organizations seeking to benefit from the Internet.

Rebecca is a graduate of Wellesley College, and received an MA from the University of Bradford in West Yorkshire, U.K., and an MBA from Babson College.

ABOUT IDG BOOKS WORLDWIDE

Welcome to the world of IDG Books Worldwide.

IDG Books Worldwide, Inc., is a subsidiary of International Data Group, the world's largest publisher of computer-related information and the leading global provider of information services on information technology. IDG was founded more than 25 years ago and now employs more than 8,500 people worldwide. IDG publishes more than 275 computer publications in over 75 countries (see listing below). More than 60 million people read one or more IDG publications each month.

Launched in 1990, IDG Books Worldwide is today the #1 publisher of best-selling computer books in the United States. We are proud to have received eight awards from the Computer Press Association in recognition of editorial excellence and three from *Computer Currents*' First Annual Readers' Choice Awards. Our best-selling *...For Dummies*® series has more than 30 million copies in print with translations in 30 languages. IDG Books Worldwide, through a joint venture with IDG's Hi-Tech Beijing, became the first U.S. publisher to publish a computer book in the People's Republic of China. In record time, IDG Books Worldwide has become the first choice for millions of readers around the world who want to learn how to better manage their businesses.

Our mission is simple: Every one of our books is designed to bring extra value and skill-building instructions to the reader. Our books are written by experts who understand and care about our readers. The knowledge base of our editorial staff comes from years of experience in publishing, education, and journalism — experience we use to produce books for the '90s. In short, we care about books, so we attract the best people. We devote special attention to details such as audience, interior design, use of icons, and illustrations. And because we use an efficient process of authoring, editing, and desktop publishing our books electronically, we can spend more time ensuring superior content and spend less time on the technicalities of making books.

You can count on our commitment to deliver high-quality books at competitive prices on topics you want to read about. At IDG Books Worldwide, we continue in the IDG tradition of delivering quality for more than 25 years. You'll find no better book on a subject than one from IDG Books Worldwide.

John Kilcullen
CEO
IDG Books Worldwide, Inc.

Steven Berkowitz
President and Publisher
IDG Books Worldwide, Inc.

Eighth Annual Computer Press Awards 1992

Ninth Annual Computer Press Awards 1993

X
WINNER

Tenth Annual Computer Press Awards 1994

Eleventh Annual Computer Press Awards 1995

IDG Books Worldwide, Inc., is a subsidiary of International Data Group, the world's largest publisher of computer-related information and the leading global provider of information services on information technology. International Data Group publishes over 275 computer publications in over 75 countries. Sixty million people read one or more International Data Group publications each month. International Data Group's publications include: **ARGENTINA:** Buyer's Guide, Computerworld Argentina, PC World Argentina; **AUSTRALIA:** Australian Macworld, Australian PC World, Australian Reseller News, Computerworld, IT Casebook, Network World, Publish, Webmaster; **AUSTRIA:** Computerwelt Osterreich, Networks Austria, PC Tip Austria; **BANGLADESH:** PC World Bangladesh; **BELARUS:** PC World Belarus; **BELGIUM:** Data News; **BRAZIL:** Annuário de Informática, Computerworld, Connections, Macworld, PC Player, PC World, Publish, Reseller News, Supergamepower; **BULGARIA:** Computerworld Bulgaria, Network World Bulgaria, PC & MacWorld Bulgaria; **CANADA:** CIO Canada, Client/Server World, ComputerWorld Canada, InfoWorld Canada, NetworkWorld Canada, WebWorld; **CHILE:** Computerworld Chile, PC World Chile; **COLOMBIA:** Computerworld Colombia, PC World Colombia; **COSTA RICA:** PC World Centro America; **THE CZECH AND SLOVAK REPUBLICS:** Computerworld Czechoslovakia, Macworld Czech Republic, PC World Czechoslovakia; **DENMARK:** Communications World Danmark, Computerworld Danmark, Macworld Danmark, PC World Danmark, Techworld Denmark; **DOMINICAN REPUBLIC:** PC World Republica Dominicana; **ECUADOR:** PC World Ecuador; **EGYPT:** Computerworld Middle East, PC World Middle East; **EL SALVADOR:** PC World Centro America; **FINLAND:** MikroPC, Tietoverkko, Tietoviikko; **FRANCE:** Distributique, Hebdo, Info PC, Le Monde Informatique, Macworld, Reseaux & Telecoms, WebMaster France; **GERMANY:** Computer Partner, Computerwoche, Computerwoche Extra, Computerwoche FOCUS, Global Online, Macwelt, PC Welt; **GREECE:** Amiga Computing, GamePro Greece, Multimedia World; **GUATEMALA:** PC World Centro America; **HONDURAS:** PC World Centro America; **HONG KONG:** Computerworld Hong Kong, PC World Hong Kong, Publish in Asia; **HUNGARY:** ABCD CD-ROM, Computerworld Szamitastechnika, Internetto online Magazine, PC World Hungary, PC-X Magazin Hungary; **ICELAND:** Tolvuheimur PC World Island; **INDIA:** Information Communications World, Information Systems Computerworld, PC World India, Publish in Asia; **INDONESIA:** InfoKomputer PC World, Komputek Computerworld, Publish in Asia; **IRELAND:** ComputerScope, PC Live!; **ISRAEL:** Macworld Israel, People & Computers/Computerworld; **ITALY:** Computerworld Italia, Macworld Italia, Networking Italia, PC World Italia; **JAPAN:** DTP World, Macworld Japan, Nikkei Personal Computing, OS/2 World Japan, SunWorld Japan, Windows NT World, Windows World Japan; **KENYA:** PC World East African; **KOREA:** Hi-Tech Information, Macworld Korea, PC World Korea; **MACEDONIA:** PC World Macedonia; **MALAYSIA:** Computerworld Malaysia, PC World Malaysia, Publish in Asia; **MALTA:** PC World Malta; **MEXICO:** Computerworld Mexico, PC World Mexico; **MYANMAR:** PC World Myanmar; **NETHERLANDS:** Computer! Totaal, LAN Internetworking Magazine, LAN World Buyers Guide, Macworld Netherlands, Net, WebWereld; **NEW ZEALAND:** Absolute Beginners Guide and Plain & Simple Series, Computer Buyer, Computer Industry Directory, Computerworld New Zealand, MTB, Network World, PC World New Zealand; **NICARAGUA:** PC World Centro America; **NORWAY:** Computerworld Norge, CW Rapport, Datamagasinet, Financial Rapport, Kursguide Norge, Macworld Norge, Multimediaworld Norge, PC World Ekspress Norge, PC World Nettverk, PC World Norge, PC World ProduktGuide Norge; **PAKISTAN:** Computerworld Pakistan; **PANAMA:** PC World Panama; **PEOPLE'S REPUBLIC OF CHINA:** China Computer Users, China Computerworld, China InfoWorld, China Telecom World Weekly, Computer & Communication, Electronic Design China, Electronics Today, Electronics Weekly, Game Software, PC World China, Popular Computer Week, Software Weekly, Software World, Telecom World; **PERU:** Computerworld Peru, PC World Profesional Peru, PC World SoHo Peru; **PHILIPPINES:** Click!, Computerworld Philippines, PC World Philippines, Publish in Asia; **POLAND:** Computerworld Poland, Computerworld Special Report Poland, Cyber, Macworld Poland, Networld Poland, PC World Komputer; **PORTUGAL:** Cerebro/PC World, Computerworld/Correio Informático, Dealer World Portugal, Mac*In/PC*In Portugal, Multimedia World; **PUERTO RICO:** PC World Puerto Rico; **ROMANIA:** Computerworld Romania, PC World Romania, Telecom Romania; **RUSSIA:** Computerworld Russia, Mir PK, Publish, Seti; **SINGAPORE:** Computerworld Singapore, PC World Singapore, Publish in Asia; **SLOVENIA:** Monitor; **SOUTH AFRICA:** Computing SA, Network World SA, Software World SA; **SPAIN:** Communicaciones World España, Computerworld España, Dealer World España, Macworld España, PC World España; **SRI LANKA:** Infolink PC World; **SWEDEN:** CAP&Design, Computer Sweden, Corporate Computing Sweden, Internetworld Sweden, it.branschen, Macworld Sweden, MaxiData Sweden, MikroDatorn, Nätverk & Kommunikation, PC World Sweden, PCaktiv, Windows World Sweden; **SWITZERLAND:** Computerworld Schweiz, Macworld Schweiz, PCtip; **TAIWAN:** Computerworld Taiwan, Macworld Taiwan, NEW ViSiON/Publish, PC World Taiwan, Windows World Taiwan; **THAILAND:** Publish in Asia, Thai Computerworld; **TURKEY:** Computerworld Turkiye, Macworld Turkiye, Network World Turkiye, PC World Turkiye; **UKRAINE:** Computerworld Kiev, Multimedia World Ukraine, PC World Ukraine; **UNITED KINGDOM:** Acorn User UK, Amiga Action UK, Amiga Computing UK, Apple Talk UK, Computing, Macworld, Parents and Computers UK, PC Advisor, PC Home, PSX Pro, The WEB; **UNITED STATES:** Cable in the Classroom, CIO Magazine, Computerworld, DOS World, Federal Computer Week, GamePro Magazine, InfoWorld, I-Way, Macworld, Network World, PC Games, PC World, Publish, Video Event, THE WEB Magazine, and WebMaster; online webzines: JavaWorld, NetscapeWorld, and SunWorld Online; **URUGUAY:** InfoWorld Uruguay; **VENEZUELA:** Computerworld Venezuela, PC World Venezuela; and **VIETNAM:** PC World Vietnam. 3/24/97

Dedication

Danny thanks his wife Holly, for her continuous, tireless support. She has put up with so many special projects and the phrase "But this is different" so many times that this book could not be done without her. (She remains curious, however, as to how a true dummy can write a book "...*For Dummies.*") Danny also wants to thank his parents for putting him through challenging English courses and an unbearable summer (yes, summer!) school typing course that made writing this book relatively easy. Finally, he thanks Pat for making the nuts and bolts of these pages work — the core "point and click" words are Pat's.

Pat also offers special thanks to his wife, Christine, who heard the phrase "I'll come to bed before two a.m. tonight — really!" way too many times while this book was being written, and who introduced him to his two favorite things besides her: Macs (the computers, not the hamburgers), and the Internet. He also thanks his parents for providing food and shelter (and much, much more) while he was in New Jersey working on this book, and for never complaining once about his being grumpy.

Rebecca thanks her mother, writer and journalist Betty Wetzel, who taught her how to write and to love it.

Authors' Acknowledgments

As with any book, the help of countless individuals made this one possible. We could never thank everyone individually, but we're certainly going to give it a try.

Geoff Graham took time away from work, school, and (more than once) sleep to get online with us while we were testing new applications and putting together our artwork. He even parted with a computer for a couple of months — perhaps the ultimate sacrifice for a Computer Engineering major. His help was invaluable, and his flexibility above and beyond the call of duty (It's amazing what you can get a college student to do for you in exchange for a hot meal.)

Thanks to Larry Fromm of Dialogic, who shared his knowledge and insight of the Internet telephony gateway market, and helped show us the future of Internet telephony.

Dozens of vendor representatives took the time to offer us tips, techniques, and critiques as we wrote about their products. Special thanks go to Adam Rauch of Microsoft, Duane Fields of Netscape, Harvey Kaufman of NetSpeak, Tracy Specht of White Pine Software, and Thomas Walsh of Lucent Technologies, and everyone else who answered our questions and read our first drafts.

Thanks to Kevin Savetz, who shared his excellent insight on video broadcasting on the Internet. Check out his book, which he coauthored with Neil Randall and Yves Lepage, *MBone: Multicasting Tomorrow's Internet* (published by IDG Books Worldwide, Inc.)

Special thanks to Denise Myers, our technical reviewer. She kept us honest, and made sure we were always telling the truth — we're sure her work will make your voyage into Internet telephony and video smoother. Thanks also to Denise's partner from Electric Magic, Andrew Green, who helped us a lot with the first edition of this book and introduced us to Denise.

Finally, thanks to the IDG team who got us going on this book and made sure that we remembered to dot our *i's* and cross our *t's*. We especially thank our project editor, John Pont, for his hard work and excellent advice.

Publisher's Acknowledgments

We're proud of this book; please send us your comments about it by using the IDG Books WorldWide Registration Card at the back of the book or by e-mailing us at `feedback/dummies@idgbooks.com`. Some of the people who helped bring this book to market include the following:

Acquisitions, Development, and Editorial

Project Editor: John Pont

Senior Acquisitions Editor: Jill Pisoni

Acquisitions Editor: Michael Kelly

Product Development Director: Mary Bednarek

Media Development Manager: Joyce Pepple

Associate Permissions Editor: Heather H. Dismore

Technical Editor: Denise M. Myers, Internet Telephony Pioneer

Editorial Manager: Mary C. Corder

Editorial Assistant: Chris Collins

Production

Project Coordinator: Shawn Aylsworth

Layout and Graphics: Cameron Booker, Linda Boyer, Drew Moore, Mark Owens, Kate Snell

Proofreaders: Arielle Carole Mennelle, Carrie Voorhis, Robert Springer, Karen York

Indexer: Ty Koontz

Special Help
Suzanne Thomas, Associate Editor; Tina Sims, Copy Editor; Stephanie Koutek, Proof Editor; Joell Smith, Associate Technical Editor; Access Technology

---

General and Administrative

IDG Books Worldwide, Inc.: John Kilcullen, CEO; Steven Berkowitz, President and Publisher

IDG Books Technology Publishing: Brenda McLaughlin, Senior Vice President and Group Publisher

Dummies Technology Press and Dummies Editorial: Diane Graves Steele, Vice President and Associate Publisher; Judith A. Taylor, Product Marketing Manager; Kristin A. Cocks, Editorial Director

Dummies Trade Press: Kathleen A. Welton, Vice President and Publisher

IDG Books Production for Dummies Press: Beth Jenkins, Production Director; Cindy L. Phipps, Manager of Project Coordination, Production Proofreading, and Indexing; Kathie S. Schutte, Supervisor of Page Layout; Shelley Lea, Supervisor of Graphics and Design; Debbie J. Gates, Production Systems Specialist; Tony Augsburger, Supervisor of Reprints and Bluelines; Leslie Popplewell, Media Archive Coordinator

Dummies Packaging and Book Design: Patti Sandez, Packaging Specialist; Lance Kayser, Packaging Assistant; Kavish + Kavish, Cover Design

♦

The publisher would like to give special thanks to Patrick J. McGovern, without whom this book would not have been possible.

♦

Contents at a Glance

Cartoons at a Glance

By Rich Tennant

page 325

"IT HAPPENED AROUND THE TIME WE SUBSCRIBED TO AN ONLINE SERVICE."

page 9

"You know, I liked you a whole lot more on the Internet."

page 275

"IT'S JUST UNTIL WE GET BACK UP ON THE INTERNET."

page 127

"WELL, I'M REALLY LOOKING FORWARD TO SEEING THIS WIRELESS E-MAIL SYSTEM OF YOURS, MUDNICK."

page 247

Fax:* 508-546-7747 • *E-mail: the5wave@tiac.net

Table of Contents

Foreword

With the invention of the telegraph, the bonds that tied speed of communication to speed of transportation were broken forever. The Pony Express, a recent startup, promptly went out of business. Somehow, though, people didn't flock to install telegraph keys in their kitchens; this primitive form of e-mail was a great success in government and big businesses but not in the home. The user interface was too complex and telegraph clicks too inexpressive.

A hundred years ago, the telephone network began to grow to connect users of Alexander Graham Bell's wonderful device. Phones, too, started out in business but quickly spread to homes; voice is simple and voice is compelling.

Forty years ago, one-way video, for better or worse, became the way the developed world is informed and entertained.

Some 25 years ago, what we now know of as the Internet was established to connect engineers and scientists in universities, government, and the defense industry. No devices existed for home use that could take advantage of the Internet for communication. What's more, access was walled off by a thicket of UNIX jargon that passed for a user interface.

Four years ago, the Mosaic browser made the content of the World Wide Web graphically and easily available. The creative chaos of hyperlinks (URLs) replaced the structured hierarchies of UNIX directories. A flood of innovation was released, and the Web has become a part of our home life as well as our business. Studies show that people are spending time on the Web that they used to spend watching television (the time had to come from somewhere; people are busy).

Internet Telephony For Dummies, 2nd Edition is about the convergence of our most familiar forms of communication — telephone, video, and fax — with the reach and graphics of the Internet and the World Wide Web. The book is about audio in general, and voice in particular, combined with graphics and augmented by fax. It is about two-way video — about being able to view the video you want when you want it. It is about the continuing trend toward less-expensive communication between any group of people, anywhere, at any time, and *in any form.*

Internet Telephony For Dummies, 2nd Edition is about the future that is beginning to happen now. The book details the software and hardware you need, and the procedures to follow, to be part of this evolving future today.

Tom Evslin, Vice President, AT&T WorldNet Service

Tom Evslin was solidly ensconced in the computing and communications industries when the Internet arrived, and is now thoroughly involved in its continuing evolution as head of AT&T WorldNet Service, with responsibility for the largest pure Internet service provider in the United States.

Introduction

Welcome to *Internet Telephony For Dummies,* 2nd Edition. This book is among the first to specifically tell you how to save money *and* do lots of nifty productivity-improving things on the Internet (which we affectionately refer to as "the 'Net") — all through the use of Internet telephony and video.

Telephony is pronounced with the accent on the *le*, as in *te-LE-phony* (and not at all like *Tell-A-Phoney*, which really annoys us.) Telephony includes the telephone, video telephones, videoconferencing, audioconferencing, collaborative conferencing, and every other type of conferencing (except actual face-to-face) that you can think of.

Telephony has traditionally revolved around the telephone networks. But the Age of the 'Net means that the computer networks now have a shot, too. The tools we use for communicating with one another are changing radically, prompting drastic changes in old habits — such as picking up the phone and placing a telephone call or wandering over to the fax machine to wait in line to send your paper documents across the globe.

By the time you finish with this book, you'll be able to make a voice call to anyone in the world from your computer, either through the Internet or the telephone networks. You discover how to use the Internet to speak, smile, primp, and argue with people similarly inclined and, in so doing, put yourself on the leading edge of a truly worldwide revolution in telephony.

We don't tell you that Internet telephony solves all your communications problems or that you can toss your telephone. But we do describe in detail how Internet telephony can handle, admirably, many of your communication needs. In many cases, you get a lot more features and functionality out of a computer network — features that you could never afford from the telephone company. Telephony is growing like wildfire on the 'Net — and this book can put you on the leading edge of this new technology.

About This Book

As with all *...For Dummies* books, this one takes a refreshing departure from the standard Internet approach, which often takes a very computer-centric tact. We designed this book to bridge two rapidly converging camps, computers and telecommunications, and to show you how you can use the former (computers) to do the stuff that you used to rely on the latter (telecommunications) to do — and save some money at the same time.

So all you computer nerds who mumble UNIX commands in your sleep (we're talking to the two or three of you who actually read *...For Dummies* books, that is) but who don't know that Ma Bell got broken up back in 1985 now have a chance to parlay that supercharged five million megahertz computer into a neat substitute for your phone and make that Internet connection really pay off.

And for all of you who spend most of your time comparing telephone providers and trying to decide who's lying the most in those TV ads for discount long-distance plans, this book provides an opportunity to find out how the most innovative and fastest-growing communications phenomenon (the Internet, in case we weren't obvious enough) can help substantially cut your costs while giving you loads of neat new features that you'd normally have to pay a bundle for, like video and Caller ID.

This book does, however, assume that you know some basic concepts about our subject, such as how to use a telephone, what a videoconference is, and that the Internet is a Valuable Thing. Otherwise, we start from scratch. And we intend to base our discussion on a perspective appropriate for both the computer and communications camps.

Pertinent sections of this book cover the following points:

- What the Internet is and how Internet telephony fits into all of this Internet mania you read about
- What you need to telecommunicate over the 'Net
- Who's got what for voice, video, and multimedia collaboration
- How to place telephone calls over the Internet
- The best way to use videoconferencing
- Common mistakes and how to correct them — or avoid them altogether
- Where to look for useful information on Internet telephony beyond what's found in this book
- What is coming down the road for Internet telephony

Our goal is to give you enough information about Internet telephony so that you can look at your present circumstances and understand what you can do on the 'Net — as well as decide whether you want to try Internet telephony at all. (We think you will.)

Oh, and we refuse to use any of the following analogies in this book:

- Riding the Information Superhighway, including references to on-ramps, off-ramps, speed limits, way stations, rest stops, lanes, traffic tie-ups, and so on.
- Surfing the 'Net, including references to waves, dudes, and bitchin' information.

As avid *Star Trek* fans, however, we reserve the right to quote from any episode from any of the series at will, without liability from the reader. And consider yourselves warned, too, that one of us likes cars even more than the Internet, so don't hold any automotive metaphors against us, okay?

If you've seen the first edition of *Internet Telephony For Dummies*, you may be wondering what's new in the second edition. Here's a short answer to that question: LOTS! Internet technology moves like lightning — and we're here to keep you caught up. Here are some of the new things you can find in this edition:

- The bottom line on new Internet telephony standards — and what they mean to you
- Step-by-step instructions for setting up and using the latest and greatest Internet telephony software
- An entire section devoted to the gateway hardware that allows regular phones and fax machines to get into Internet telephony

Where to Go from Here?

The rest of this introduction focuses on helping lay out the organization of this book, so that you know what to expect along the way. The book itself consists of four main parts.

Part I: Internet Telephony and Video, Unplugged

In the first part of the book (we don't call it Part I for nothing!), we explain a little about the Internet and how it works. No histories — you've heard all that before — but rather some techie stuff (not too techie, though — this *is* a *...For Dummies* book, after all) that tells you what's really going on whenever you make a phone call or get involved in a videoconference on the 'Net. Why do you need to know this information? Not just to impress your friends or boss, but so that you have a better idea of what you can and can't do on the 'Net. We also give you a broad overview of just about everything that's being used on the 'Net right now — what programs are available and what they can do.

Part II: You Make the Call: Using Internet Telephony Products

In Part II, we tell you — in detail — how to use five of the most popular telephony programs on the 'Net. If you're like us (and we bet that you are) you're going to want to get online and start talking right away.

Part III: Internet Telephony Not Just for PCs Anymore

The biggest change in the world of Internet telephony since we wrote the first edition of *Internet Telephony For Dummies* last year has been the development of gateways and gateway services. These systems let regular telephones and fax machines hook into the Internet and communicate with computers, or even other regular phones and faxes over the 'Net. In Part III, we let you know what's happening with this merger of the telephone network and the Internet, and how you can use some of these products in your home or business.

Part IV: The Part of Tens: Hip Stuff from the Internet

We don't write for Letterman, so don't expect these top-ten lists to be *funny,* but some of them are *fun.* Some are more plainly informative, but — hey! — knowledge is its own reward, right?

Part V: Appendixes

Appendix A helps you track down an Internet service provider, and Appendix B helps you track down some of the companies that sell the Internet telephony gateway products we discuss in Part III of the book. Appendix C gets you started using the programs included on the CD-ROM (which is tucked into the back cover of this book). We suggest that you read at least part of this book before you fire that baby up, but if you don't do that, at least peruse Appendix C for installation instructions for the CD and for the programs on the CD.

Icons Used in This Book

We'd be surprised if this were the first ...*For Dummies* book that you've ever read. (If so, shame, shame, shame.) But in case it is, we want to mention the icons we use to encourage you to pause at particular points in the book.

Flags good advice, including shortcuts and time-saving secrets.

Points out extra neat stuff that makes this technology actually work.

Marks information you need to keep in mind.

Flags areas that are potentially troublesome.

Highlights groovy information that you may not need but might just like to know.

Conventions Used in This Book

To be as consistent and helpful as possible, we adopted a few conventions.

In several places, we provide you with a series of steps to complete in a certain order. Be sure to follow these steps in the order specified. They look something like the following:

1. **Click on the Joined Chat Rooms button.**
2. **Click on the name of the Chat Room you want to leave in the Chat Room list.**

The Leave/Join button toggles to Leave.

3. **Click on the Leave button.**

   You're outta there.

In other places, we provide you with bulleted lists.

- Some of these lists provide steps that you can complete in any order.
- Others provide descriptive information.

We type the entries that we ask you to make on-screen in **boldface** (unless they are in a numbered step, in which case they appear in a normal typeface).

Keyboard sequences where you press and hold the first key and then press the second are written with a plus sign, like Alt+X (for pressing and holding the Alt key and then pressing the X key).

If we ask you to select something from a menu, we often use the command arrow to indicate that you click on the first item and then select the second item from the menu. For example, if we say "choose File⇨Exit," we mean click on the File menu with your mouse and then scroll down and select Exit. ***Remember:*** On a Macintosh, you must continue pressing the mouse while doing this step; Windows allows you to click once to open the menu and then a second time to select the second item.

Underlined letters in commands from menus and dialog boxes (like the F in File above) indicate hot keys or keyboard shortcuts. Press and hold the Alt key on a Windows computer, or the Command (Apple) key on a Mac, and then press this key on your keyboard to perform the operation that all that mouse wrangling performs.

We also give you World Wide Web addresses (URLs, or Uniform Resource Locators) to access with your Web browser. You can use these addresses, which are typed in monofont (for example, `http://www.vocaltec.com`), to find software or information (or just for fun sometimes). Just type these addresses exactly as they're shown in your Web browser's Goto window, and you'll get there.

Finally, we provide words in *italics* to indicate the first time that we use a new term or acronym.

Feedback, Please

We close with a plea for you to be interactive. We'd love to hear where in this tome we could have been more clear, less clear, more specific, less digressive, and so on. If you want to contact us, please feel free to do so at the following address:

TeleChoice, Inc.
15 Bloomfield Ave., Suite 3
Verona, NJ 07044

Or send us an e-mail message at `dummies@telechoice.com`.

Things are changing fast on the 'Net, and changes occur even faster in the world of Internet telephony and video. We found ourselves going back and rewriting or adding new chapters the entire time we wrote this book (even after, as a few new products almost made us yell "stop the presses"). We want you to keep up to date — so for continual updates, news, and reviews on Internet telephony and video software, we invite you to point your browser to our Web site, `http://www.telechoice.com`.

Part I

Internet Telephony and Video, Unplugged

"IT HAPPENED AROUND THE TIME WE SUBSCRIBED TO AN ONLINE SERVICE."

In this part . . .

In the olden days (back about 1994), the Internet was used for sending e-mail messages, transferring files, and, on occasion, looking at Web pages. Then the Web hit the big time, and the 'Net became a multimedia event.

Pretty soon everyone was on the Web, seeing what it could do and clamoring for more. That's where Internet telephony and video come in — they make the multimedia part of the 'Net come alive. In this part, we tell you what you *want* and *need* to know about how the Internet and Internet telephony work.

We also tell you which telephony and video programs are available for your computer, whether Mac or Windows, and we show you what to look for if you aren't connected to an Internet Service Provider.

Chapter 1

What's Internet Telephony and Video?

In This Chapter

- What's all the hubbub?
- Phone home — over the 'Net!
- Using your computer to make voice and video calls
- Taking advantage of fantastic calling features
- Talking to friends around the world — free!

If you think the Internet is cool because you can get instant information on just about any topic imaginable, wait until you use speech and video over the Internet. It's a whole new world.

Internet telephony and video are on the cusp of changing the way we interact on a daily basis. It's not that using a telephone is new (neither is videoconferencing for that matter). But applying them to the easy-to-use, ubiquitous Internet *is* new and, frankly, quite state-of-the-art.

From the comfort of your favorite computer chair, you can use Internet telephony to call long lost friends via an Internet telephone or to check out your brother's kids as they get all dressed up for Halloween. Internet telephony and video also let you interface with suppliers, vendors, consultants, advisors, information providers, news organizations, and others in totally new ways.

Internet telephony and video are changing the nature of communications as we know it. Is this all hype? Well, you will just have to see for yourself. This book tells you how to jump right into the fray.

Sending Your Voice and Video over the Internet

Essentially, using *Internet telephony and video* means using the Internet to make telephone and video calls. All Internet telephony and video products are *streaming* media, which means that the audio or video starts playing back as soon as it is received — the beginning plays while the end is still on its way. This is different from traditional media on the Internet, where the entire file is downloaded and then played back.

Here are some things you can do:

- **One-on-one telephony:** You can place a voice call to anyone on the Internet. The process can be as simple as clicking on someone's name, e-mail address, or *Internet Protocol* (IP) address.

  Some companies are even offering services that let you call regular telephones as well — like the phone in your kitchen or the one on the desk in your office.

- **One-on-one videoconferencing:** You can make a multimedia video call to anyone who has the requisite hardware and software. This means that you can see the person you are talking to, and the more sophisticated programs even allow you to share whiteboards (basically electronic versions of the whiteboards on the wall of your office or classroom) and documents for group editing. Internet video programs are especially popular with groups of people who work on team projects from multiple locations.

- **Audio on demand:** Audio on demand allows you to play stored audio-clip files in real time (sometimes dubbed *real-time audio*). You click on an icon or picture representing the clip, and the sound plays across your Internet connection — like listening to a tape recorder. Because clips exist as files stored either on the host computer or your own computer, you can listen to them over and over again. Increasingly, audio on demand lets you hear parts of speeches and listen to Welcome messages on Web pages. For example, when you log onto `http://www.polygram.com/polygram`, you can listen to clips of the newest releases from Polygram artists like the Gin Blossoms.

- **Live audio streams:** You can get live audio streams in real time, referred to as *Internet radio* or *multicast audio.* Instead of originating in a file, this transmission is live, and the sound just streams onto your computer — it's the Internet equivalent of live radio. Many radio stations now offer these live audio transmissions. The Web site for WCBS News Radio 88 (`www.newsradio88.com`) in New York City, for example, sends out news around the clock — the very same news you hear on your car radio.

- **Video on demand:** Video on demand is the multimedia version of audio on demand — *real-time video*. The concept is the same: You click on a link to the file, and the file is sent over the Internet to your computer, where the file is decompressed and played for your enjoyment. It's like hitting the play button on your VCR. Video enhances the pages of Sony Music (`http://www.sony.com/Music/VideoStuff/VideoClips/`), where you can download and view pieces of Sony's newest videos.
- **Live video streams:** And finally, you can enjoy live video — so-called *Internet TV* or *live video multicast*. Live video is television over the Internet. A person can set up a camera, turn it on, and stream the signal live to anyone who tunes in — all over the Internet. For example, you can see NBC Pro, with live news from NBC and its business network CNBC, at `http://www.xingtech.com/nbc.html`.

  Live video is the part of the whole equation that scares the TV broadcast industry! Although it's unlikely that you will start watching *Friends* or *Saturday Night Live* over the Internet, it's just a matter of time until small colleges and high schools will be able to film live broadcasts of their football games, for example. Frankly, the technology will take a while to catch up to the desires of the people who want to use it; the quality of Internet TV options is not good enough for something as fast paced as a sporting event. Nonetheless, as Internet access speeds increase, the potential becomes very real. Internet TV over a high-speed connection can approach VCR quality and very much rival a cable TV or direct broadcast satellite option.

Who can use it?

The great thing about Internet telephone and video is that anyone with the right computer can use it. It is

- **Inexpensive.** Users can download most of today's Internet telephony software directly off the 'Net, and much of this software is free. Even a fully outfitted video connection can cost less than $100 — and that includes the camera!
- **Ubiquitous.** The Internet was conceptualized by its university and military architects to go just about everywhere, and today it does. Consequently, Internet access is available around the world, and getting an account is usually as simple as providing a credit card number over the phone and downloading some software. We tell you whom to call in Appendix A!

  You can use your Internet account at work, at home, or on the road. One of the authors of this book even used his account from Snowmass Village, Colorado, alongside the ski slopes while working on this book.

Your IP address explained

Every computer that is connected directly to the Internet (by a dedicated connection or certain dial-up connections, as we discuss in Chapter 3) has a unique name on the 'Net. This name is known as the computer's IP (Internet Protocol) address, which is a series of four numbers between 1 and 254 and looks something like 111.222.111.222. This number identifies your machine's location on the 'Net and allows every other computer on the 'Net to communicate with it (if you allow them access). You'll hear lots about IP addresses as we continue — about what they are and how to get one — because without one (and you can be on the Internet — sort of — without one) you can't use most Internet phone and video programs.

Have computer, will travel. Most new computers come outfitted with the equipment you need to use Internet telephony: built-in microphones and sound programs. So however you use a computer, you can take advantage of this new technology.

- **Nondiscriminatory:** The Internet is available to everyone, and as such, everyone can use Internet telephony. This is not a business-only or residential-only type of service. Although the ante of a multimedia-style computer could limit the accessibility of Internet telephony and video, prices continue to drop for computers, making them affordable for more and more people. This universal access means that moms, dads, grandmas, grandpas, college students, kids, and robots — yes, even computers themselves — are logging on and talking with people over the Internet.

Just whom may I call?

You may call practically anyone you want to. You can

- **Talk to people you know.** Set up a prearranged time to talk with your friends, or if you have full-time Internet access, such as in a business setting, you can merely leave your program on and let people dial in whenever they want. A pop-up window and ring let you know someone's looking for you.
- **Talk to people you don't know.** Users of Internet telephony and video programs want to interact with people they don't even know. The Internet is the ham radio of the 1990s! Log on, click on a stranger's name, and see what you get.

- **Log onto audio and video servers.** Log onto computers that store or stream audio and video. Watch Bill Gates's speech to a computer conference, tune into your home radio station, watch a fish tank, or hear Fred Fishkin's computer and software recommendations in WCBS Radio's daily computer tutorial, *Bootcamp,* in New York City. Some servers on the Internet allow multiple people to join a conversation at once. With some software, people can share files, write together on the same whiteboard, and talk with or view each other, all at the same time. You can find Bootcamp on the Web at `http://cgi.pathfinder.com/netly/bootcamp/`.

Where Is It All Going?

Soon you'll be able to call anyone in the world — on telephones, fax machines, and other standard telephony devices — through Internet interfaces with the telephone network. You will not have to limit your conversations to people who happen to be on the Internet at a particular time. You can try to reach someone's computer first, and failing to connect there, your program will try that person's home phone, and then cellular phone, and then pager, and so on. This busywork will all be controlled by your software and interactions with sophisticated Internet services.

So if you are traveling along the road and need to call your associate at his computer, you can dial a toll-free number and punch in the address you want, and a computer starts ringing in your associate's office.

At least this is what the telephony companies are promising at this time. Here is some proof that they are making progress:

- **MCI's VAULT:** MCI's recently announced new network architecture combines the company's two (currently separate) networks: the MCI conventional telephone network and its substantial Internet network. (MCI provides a significant chunk of the Internet's *backbone* — the telecommunications network that links all the networks that make up the 'Net.) With VAULT, MCI customers can do things like check e-mail and make voice calls, or view a Web page and talk to a customer service agent from one single connection.
- **The Lucent Internet Telephony Server:** This newly announced product from Lucent (the telecommunications equipment manufacturer that was recently spun-off from AT&T) is the latest in the growing family of *gateway* products — computers that connect traditional telephones and fax machines to the Internet, using Internet telephony technologies to send voice, video, or even fax signals without using traditional telephone carriers.

Why Do It over the Internet?

It is very possible that the Internet will soon challenge the regular telephone network as a way to place your telephone calls to any phone anywhere in the world. Using Internet telephone and video software has many advantages over using the regular telephone network to make the phone calls:

- **Save money!** You do not have to subscribe to a special service to make Internet telephony calls, so you can use the same service for surfing the 'Net as you use for making telephone or video calls. You can find unlimited-use Internet access services for less than $20 a month. Many of you reading this book may already use these providers for surfing the 'Net, so telephone calling is basically a free add-on for you. This add-on can save you big bucks. Imagine that you are using a long-distance service that charges 10 cents per minute to call anywhere in the United States. As you see in Figure 1-1, at just over three hours of use on a $19.95 unlimited-use Internet access service, you break even — and get Internet surfing for free.
- **Get more out of your phone.** Internet telephony and video allow you to take the first steps toward using your phone in some very sophisticated ways. For example, the latest trend in telephones — computer telephony integration (CTI) — allows you to use your computer to control your telephone. In this case, because your computer is your telephone, you can take advantage of CTI features without having to buy a new telephone. What do you get with CTI? Nifty features for one thing. Look at these:
    - **Caller ID:** You can find out who is calling before you answer the phone (and avoid those people you'd rather not talk to).

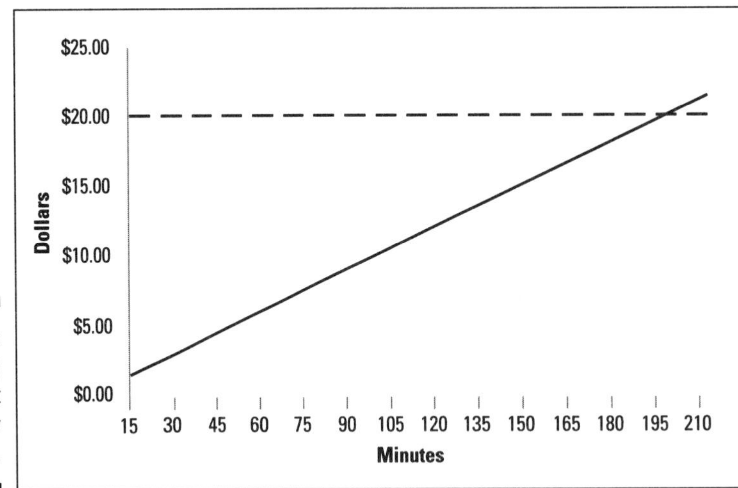

Figure 1-1: Making Internet telephony pay off.

IP and e-mail addresses: apples and oranges

So you have an e-mail address. You know it, your friends know it, and sometimes people who want to send you junk mail know it (don't you hate that!). It's always the same, no matter where or how you connect to the Internet, and your mail always finds you. That's because your e-mail doesn't usually go directly to your computer, but rather to a computer called a mail server that's on a fixed location (a nonchanging IP address) on the Internet. When you log onto the Internet and start up your mail program, your computer basically tells the mail server, "Hey, here I am, give me my mail!"

Your e-mail address is really just a proxy or alias for a location on your mail server; it doesn't identify the location of your own computer. Your IP address, on the other hand, defines the actual location of your computer on the Internet — and it can, and does, change when, for example, you dial in to your Internet Service Provider (which usually assigns you a different IP address every time you connect).

- **Call notification from the background:** As long as you are online (or all of the time if you have dedicated access), you can run a minimized or background application, do other work or explore the Web, and be notified of a call from the background by a pop-up screen or customized sound. Or if you want, you can configure your program to answer your call or videoconference automatically.
- **Call notification by e-mail:** Want to call someone who's not online? You can configure your program to automatically send that person an e-mail message saying "I called and you weren't in."
- **Dial to an e-mail address:** Forget memorizing IP addresses. Programs have the flexibility to use alternative addresses, like e-mail addresses, instead. Most people's IP address changes every time they log onto the 'Net (see Chapter 3 for more on this), while e-mail addresses stay the same — so they're a whole lot easier to remember.
- **Call forwarding:** Going down the hall for a few hours? Forward your phone to another IP address and receive your calls there.
- **Call holding:** Just as on your normal telephone, some programs can put multiple calls on hold at one time.
- **Voice messaging:** Use your computer as your own answering machine. Powerful voice messaging capabilities allow you to answer calls when you are not available.
- **Record/play back calls:** With your own recording system at your fingertips, forget about expensive telephone recording machines. Your Internet telephone can record the call so that you can play it back later. Moreover, you can use sound studio programs to clip and even store recordings in neat places, like on your personal Web page if you have one.

- **Encryption:** Concerned about someone on the Internet snooping in on your phone calls? New encryption options allow you to code/decode your calls so that only those people who are supposed to hear what is being said can do so.
- **Conference calls:** Want to talk to several people at once? Internet telephony software has been specifically designed to allow multiple people to join the conversation — something that is hard to do with a simple telephone. With Internet telephony software, you can start a conference call with just a click of a button. With the right video options, you can log onto special servers to participate in video conference calls, too.

All of these features combine to create a super telephone on your desk, one that would be very expensive to replace with regular telephone options.

- **Get more out of your computer.** Internet telephony and video allow you to finally unlock your computer's multimedia potential. Many people buy advanced computing equipment with all sorts of audio/visual add-ons but are disappointed that nothing can take advantage of this sophisticated capability.

  Search no more. Internet telephony and video programs have all sorts of options to allow you to manipulate the way you interact with others over the Internet. Moreover, you can use these programs to collaborate with others via your desktop computer — without having to travel cross-country or pay a fortune in overnight express charges.

- **Visit interesting places and see interesting people without leaving your house.** This is true: One of the authors was working on a late deadline for this book, and his wife was waiting for him in Vail at her winter brain-researcher's conference (rocking event!). She was telling him how nice it was, how it was snowing, and so on. He logged onto the Vail Tourist Bureau (`http://vail.net/internetworks/whats_new/camera.big.html`) and pulled up a video picture of the area. Sure enough, light snow, beautiful sunrise, very depressing. Video can transport you to places around the globe and let you see people you may not otherwise see. Our firm has offices around the world. We would never see some of our coworkers if not for video.

- **Make the static Internet into a real-time experience.** Want to hear that speech on neurochemistry, a discussion on butterflies in Brazil, or a concert in France? The Internet has no geographical boundaries, and sending voice or video via the Internet for others to see and hear allows anyone anywhere to share your experiences. No longer is the Internet a still-picture and flat-file experience. You can go out and experience the Internet for all it's worth.

Using your telephone will never be the same after you have logged on, spoken with, interacted with, and seen your friends, colleagues, and new associates over the Internet. When you finish this book, you'll be ready to take your Internet experience to today's highest level. So hold on. . . .

Chapter 2
Internet Telephony 101

In This Chapter

- Avoiding bandwidth bottlenecks
- Understanding the limitations of Internet telephony
- Getting on the Internet
- Getting up to speed

Internet is short for *internetwork,* a term that means connecting different networks together so that they can function, everyone hopes, as one. Indeed, that's what the Internet really is — a massive network of networks that looks like a single system to you, the user.

In this book, we don't feel the urge to put forth the obligatory Internet history lesson (ARPANET, CERN, and all that stuff), because the chances are good that you've heard that about a MILLION times already — and if you haven't, you really don't need to know what happened 15 years ago to take advantage of Internet telephony and video applications today.

If you want to know where the Internet came from, how it evolved, and how to master it, check out some of the other *...For Dummies* books, such as *The Internet For Dummies,* 4th Edition, by John Levine, Carol Baroudi, and Margaret Levine Young; or *The Internet For Macs For Dummies,* 2nd Edition, by Charles Seiter.

For the purposes of this book, the Internet is a network of networks that enables end users to transmit and receive audio and video communications across a variety of interconnections. Your connection into the Internet enables you to exchange telephony and video communications with others who have similar connections (see Figure 2-1).

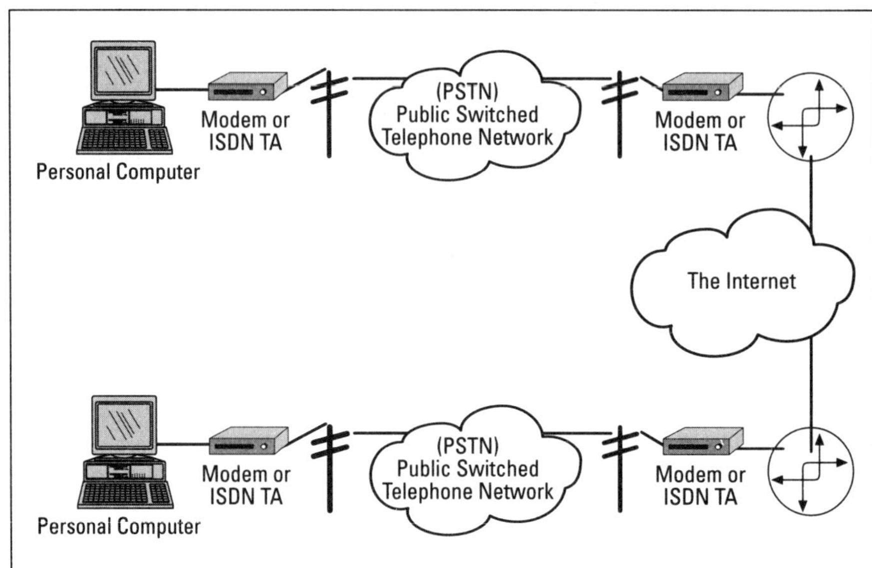

Figure 2-1: Your Internet connection.

How the Heck Do I Communicate over the 'Net?

When we talk about communicating over the Internet, we don't mean just talking or viewing video. Everything about the Internet has to do with communications, whether you're sending e-mail, sharing a file or data, or making a phone call. To the Internet, all these activities are really about the same thing — transporting bits. The Internet doesn't know whether the *packets* (small chunks of data) it transports contain your voice or someone's doctoral dissertation on fruit flies with big ears. Nor does the Internet care, generally speaking.

The Internet uses standard *protocols* to control the flow of data from one point to another. The main protocol, on which everything is based, is *Internet Protocol,* known as *IP.* (Make sure that you always remember your IP address — and remember, too, that every computer on the Internet has its own unique address that identifies that machine — and only that machine.) The switches, or *routers,* direct traffic across the 'Net and rely on the Internet Protocol to tell them where packets of data are going. Internet Protocol is really the low-level common denominator of the Internet.

Transport protocols monitor and control data transmission. The two important transport protocols are *Transmission Control Protocol (TCP)* and *User Datagram Protocol (UDP).* Internet telephony and video programs use one or the other of these protocols, and knowing more about them is important when you're ready to select your telephony software and an *Internet Service Provider (ISP).*

Transmission control protocol (TCP)

Don't worry too much about remembering the full name of this protocol; you won't be quizzed. The main thing to remember about TCP is that it's a more reliable protocol than UDP (which we describe in the following section). Data transferred by TCP is checked at the far end to ensure that all packets are received — and if some are missing or lost, TCP resends them. This protocol is great for transferring a data file or an application that doesn't work if even a small piece of data is missing. However, TCP is not so great for transmitting a phone call, because you must wait for TCP to confirm a packet's arrival and then resend the data if it doesn't show up.

User datagram protocol (UDP)

Data sent under UDP is not checked at the receiving end to ensure that all the packets have arrived, and no method exists for resending the data if some is lost along the way. Thus UDP translates into faster transfers at the expense of possibly losing some of the data. Internet telephony and video program designers usually use this system because they've decided that the increased speed and decreased delay outweigh the potential downside of losing a few data packets. You may want to check out the sidebar "Watch out for firewalls" in Chapter 3 for information about UDP and firewalls.

If you have a choice, UDP is the preferred protocol for Internet telephony and video programs.

A new addition: Real-Time Transport Protocol (RTP)

As we mention in the preceding section, developers of Internet telephony and video programs have traditionally used the UDP as the means for sending your voice and video across the 'Net. They've also realized that UDP wasn't really designed for real-time communications — it's just better than the TCP alternative. So to make things work a bit better, they've designed a new protocol — actually a pair of protocols: *Real-Time Transport Protocol (RTP)* and *Real-Time Transport Control Protocol (RTCP).*

Yet another protocol

We don't want to load you up with too much information on protocols and standards and all that underlying techie stuff that makes 'Net telephony work. After all, if you end up scratching your head and staring glassy-eyed at the book, you'll never get to the fun stuff (actually using the programs). However, we do need to mention one last protocol that is closely related to RTP. *Real-Time Streaming Protocol,* or *RTSP,* is a variant of RTP designed specifically for one-way "streaming" multimedia programs — the Internet TV and radio programs that we talk about in more detail in Chapter 8.

RTSP was developed by Netscape and some of its technology partners as a way to improve the quality of these streaming products and is now a standard. Like RTP, RTSP adds some information to the packets of audio and video data sent over the 'Net that makes it easier for programs to reproduce the multimedia when it gets to your computer.

RTP and RTCP (which work together, and are usually referred to as just RTP) help make your Internet calls work better by adding some special information to the packets you send across the 'Net. Most importantly, RTP adds special timing information to the packets to help programs get these chunks of your call data back together in order on the other end.

Basically, a *real-time communication* over the Internet is one in which the data (your voice or video) is played back on the other end of the connection immediately as you send it. In the real world, communications over the Internet (or any network for that matter) are never perfectly real-time, due to *latency.* Latency is the delay imposed upon the signal by the components of the network (it takes a bit of time for all of those routers to figure out where to send the data) and by physics (your data can only go as fast as the speed of light — which is pretty darn fast but adds a small element of delay when you're talking to someone really far away).

Note: RTP isn't actually a replacement for one of the basic Internet protocols (namely TCP or UDP). In fact, most programs that use RTP use it in addition to UDP (that is, your RTP call is sent over a UDP "connection") — so even if you communicate with a program that uses RTP, you probably also use UDP as your underlying protocol, with the same benefits and drawbacks.

Moving your data from here to there

The actual transmission of a packet across the Internet is akin to passing a baton in a relay race. A packet is sent from one switch to another. Each switch examines the packet and sends the packet where that switch thinks best. With an international transmission, having a data packet pass through tens of switches is not at all unusual.

Did you RSVP?

In the near future, you may be able to forget everything that you know about how communications on the 'Net work. That's because some major equipment vendors (Intel and Cisco Systems) and an important Internet Service Provider (BBN Planet — one of the original builders of the Internet) have just announced that they are supporting a new Internet protocol called RSVP, or Resource Reservation Protocol, which has been proposed by the Internet Engineering Task Force (one of the groups that controls how things work on the 'Net).

The RSVP protocol is a standard that enables applications, such as videoconferencing, to communicate with the network equipment in their call path and ensure that enough bandwidth is available for the duration of the session (128 Kbps in the case of the applications we are discussing here). If that level of bandwidth is available, it is reserved for the application; if not, the call isn't connected, sort of like a network busy signal.

This reservation process helps ensure that the delays, lost packets, and outright disconnects that sometimes occur today become a thing of the past. We can't wait!

The Internet uses something called *dynamic routing* to move data from one place to another. In dynamic routing, the routers that switch packets of data between points on the 'Net try to determine the best route for each individual packet as that packet comes through. The chosen route may not always be the most physically direct route. If, for example, the shortest route is tied up with traffic or has a system failure or is in the path of a major natural disaster, the Internet finds another path and sends the data across it.

Understanding general Internet issues

Being a massive network of networks, the Internet has its own set of hardware, software, protocols, procedures, and overall capabilities. Among the issues that affect how Internet telephony and video signals are carried across the network are the following:

- Bandwidth
- Packetization
- Prioritization
- Quality assurance
- Security
- Cost
- Freedom of speech

Bandwidth and compression

If you want to pump water from one place to another, how quickly you can do so depends on the size of the pipe and the speed of your pump. The Internet works the same way — the pipes are the telecommunications circuits that comprise the various networks that form the Internet.

Bandwidth refers to the amount of data a given connection (such as a modem connection) can pass in a given amount of time. Bandwidth is typically given as the number of data bits that can be sent per second, such as 28,800 bits per second. (A *bit* is the smallest unit of data that computers handle — you can think of it as the computer equivalent of an atom.) The more information contained in a signal, the bigger the pipe — that is, the more bandwidth — that's required to send the signal through in the same amount of time. The more bits per second you can fit in that pipe, the more you can send across the wire. Conversely, the more information a signal contains, the bigger the pipe required to send all of the signal in the same amount of time. As shown in Figure 2-2, the amount of bandwidth required for most audio and video signals is greater than the size of your pipe.

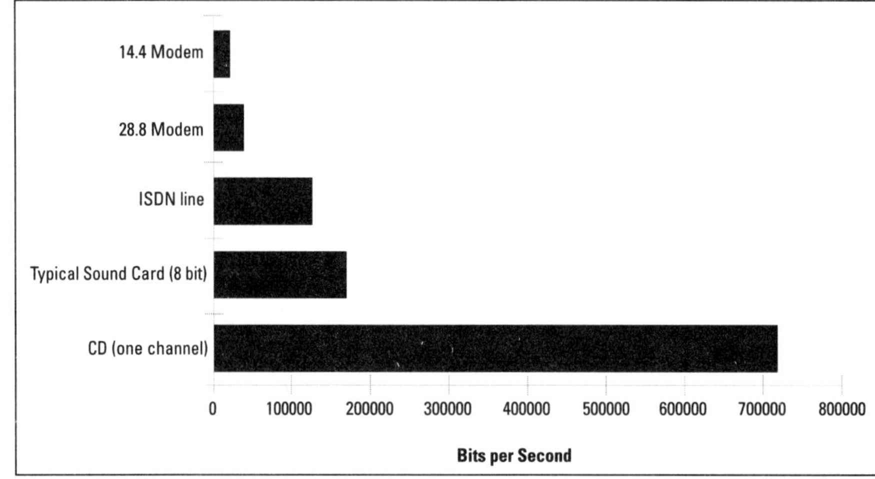

Figure 2-2: The bandwidth shortfall.

Bandwidth can obviously limit the quality of Internet telephony and video because these communications applications are time dependent. That is, things must happen within a certain time frame to make the communications usable. A minimum interval separates the bits moving over the 'Net.

Most people have a pipe of limited size, however, so the answer is to use *compression* to reduce the size of the digital signal heading through the modem or other connection point. Compression is a way of taking certain information — for example, repeated or nonessential data — out of a data stream. See the sidebar "Compression schemes (and data streams)" for more information.

The trick is to move things fast enough to meet the minimum time intervals required without sacrificing the overall data message. You must compromise somewhere, and this compromise usually rears its head in the form of *sampling* and *compression*, as defined in the following list:

- **Sampling:** How often you measure the analog voice signals to digitally store a copy of the signals for transmission.

  As an example of sampling, think of the way a movie looks on your TV when played from a VCR tape. Now think of how the same movie looked at the theater. The VCR version should seem a little less crisp in your mind and maybe not as colorful. This is because, to fit that entire movie onto a video tape, some shortcuts must be taken. The movie is *sampled* into a smaller version of itself. In most cases, however, the movie comes across just as good to the viewer as the original. Internet telephony and video work the same way to get through the bottlenecks in the network fast enough to maintain the quality of the conversation's timing.

- **Compression:** A compressed signal, using much less bandwidth, is also more Internet-friendly and less likely to clog up what is already a busy network. See the sidebar "Compression schemes (and data streams)" for more information.

Two potential bandwidth (pipe) bottlenecks can affect you:

- **In the local connection, at either end of the connection:** This connection is usually the local telephone line to which you attach some sort of modem or other transmission device.
- **On the backbone of the Internet itself:** Data flow across the Internet and through the various network interconnection interfaces can be impeded by traffic congestion, limited circuit size, and limited hardware connections, among other things.

You have control over your local connection. Choosing a faster connection method — such as a faster modem or maybe an integrated services digital network (ISDN) line — can help alleviate bandwidth limitations in your local connection, as can choosing a good Internet Service Provider (ISP) who is outfitted for this sort of Internet traffic. We help you understand how to do both items in Chapter 3.

The Internet network bandwidth limitations are, of course, out of your control but can affect the quality of your calls. You no doubt are going to notice that Web surfing during the middle of the business day, when 'Net usage is heavy, may not be as quick as it is later in the evening. That's because everyone else is trying to surf at those high peak times. Because Internet telephony and video signals are just like those of any other data flowing over the Internet, they are subject to the same congestion and quality effects at high-usage times.

Compression schemes (and data streams)

A decent compression algorithm makes or breaks an Internet telephony or video product. Many different compression schemes are available. Some, such as the GSM (Global System for Mobile Communications) algorithm used in mobile phones, are free for all to use, while others are proprietary. (Many Internet phone programs use a proprietary algorithm and are likely to advertise it as "the best.")

All compression schemes, however, fall into one of two main categories: *lossless* and *lossy*:

- Lossless compression generally finds repeated parts of the signal, removes them, and then puts them back in at the other end. The result is a signal that ends up being identical to what went into the front end of the system. This feature is very important in data applications, as in transferring Excel files, but less so for voice.
- Lossy compression methods, in comparison, use advanced modeling techniques to determine what is really important to the signal being understood and what can be left out. The signal received on the far end of this system is not the same as what went in, but if the programmers and scientists did their job, the signal should sound or look pretty much the same at both ends. These lossy compression methods, because they can remove more data, are typically more efficient (that is, they compress the signal to a greater degree) and are used for all the audio and video applications that we discuss in this book.

Some compression programs reach a 50:1 compression ratio, which means that if you have a 50MB file going in, you have a 1MB file coming out. That's pretty impressive compression.

Of course, if a Web site takes an extra 20 seconds to load, you are perturbed but not put out. With time-sensitive telephony applications, however, you can be stopped in your tracks.

Note: We're of the "if they come, someone will build it" school of thought concerning growth of Internet capacity. The almost overwhelming commercial and public demand to be online is sure to drive those groups who collectively continue to build the Internet to expand the backbone bandwidth to accommodate increasing Internet demands.

Packetization

If you look at what is happening as information streams off your computer, over the local access line, through your Internet Service Provider's switch, and into the Internet, you see that the data stream goes through some big changes. Your computer breaks down the data it sends over the Internet into smaller pieces of data known as *packets*. Each packet has a header that identifies it and its destination. Switches on the 'Net, known as *routers*, read this header, determine where the packet is going, and send it farther along on its way.

What's an Internet Service Provider?

An Internet Service Provider, or ISP, is just what the name says it is: an organization that provides Internet service or access. ISPs range from small local companies with one access number and 20 or so modems up to huge multinational corporations that provide service to subscribers all around the world.

An ISP doesn't need to be a company either. If you are connected to the Internet through a college, for example, the school is your ISP. (Although it may, in fact, be leasing its 'Net access from one of the large ISPs.)

Basically, you can think of an ISP as your on- and off-ramp to the Internet.

We talk about things to look for when choosing an ISP, and how that choice can affect your use of Internet telephony and video products, in Chapter 3.

After the packets arrive at their destination, the receiving computer reads the headers of the individual packets, puts them back in order if need be, and directs them to whatever program on your computer needs them.

Each packet could take a totally different route through the Internet to get from point A to point B. Data is sent out, bounces from place to place in a somewhat organized fashion, and eventually arrives at its destination. You have no guarantee, nor do you need one, that your next chunk of data is going to follow the same route as the previous one. This type of routing is called *connectionless* routing, because it does not depend on a specific path for the transmission.

The effect of packetization and connectionless routing is one of timing. Because successive packets in a transmission can travel different routes between two points on the Internet, they can arrive at their destination at different times. Some packets in a transmission may take a long time to arrive at their destination. The computer at the terminating end may wait and store the packets in a buffer so as to put them in correct order without dropping any packets. Some computer applications proceed without the delayed packets and send the data to the screen or output device with holes in the data stream. You may not notice these holes, because the amount of data in each packet is so small that one dropped packet is no big deal — the holes may result only in a small clipping of speech or a few portions of the screen freezing for a subsecond.

The packet arrival rate is not necessarily always staggered, however. At times when the network is underutilized, packets can arrive in their correct order and in perfect synchronization.

We call the environment in which packets may or may not arrive on time and in sequence *variable real-time communications.*

Prioritization

Prioritization — or lack thereof — is another facet of the Internet that may affect your use of Internet telephony and video programs.

Consider what happens if you make a phone or videoconference call today over the regular telephone network (which is called the *P*ublic *S*witched *T*elephone *N*etwork, or *PSTN*). Like the Internet, the PSTN is a network of networks, made up of interconnections between the long-distance carriers, local telephone companies, cellular companies, and others. The PSTN, however, is designed to offer a higher level of quality at a correspondingly higher price.

Say that you place a telephone call from your home in Peoria to your brother's house in Missoula using your carrier, MCI. After your brother answers the phone, you have a connection that is exclusively yours, running through the telephone network. You can engage in a reliable, fixed-bandwidth, stable conversation over the network, regardless of how many other people choose to place calls in the midst of your conversation.

Purists may point out situations in which this scenario is not wholly true, such as with international links, where bandwidth is constrained; in these cases, bandwidth can be scaled back as more people come onto the connection, depending on the telephone carrier.

With the Internet, the picture is slightly different. You do not have an end-to-end connection reserved for you during your communications, nor a guaranteed level of bandwidth; instead, your data travels over a shared backbone that can cause you problems during periods of high Internet usage. Internet telephony does not offer a way to achieve a higher level of quality or to obtain any sense of prioritization over other people's traffic. But the Internet is a low-cost, viable option whenever you are willing to accept its terms and limitations.

The simple solution is to call over the PSTN if your business is of a critical nature. With the PSTN, you essentially reserve your bandwidth any time you make a phone call.

Quality assurance

A related, yet slightly different, angle on the Internet is *quality control*. Because the Internet is made up of hundreds of networks and providers, you have no promise of consistency.

Consider again your call from Peoria to Missoula, except now you get cut off before you finish. If you're using a traditional local or long-distance carrier, you can contact the carriers, complain, and probably get your money back. MCI, for its part, can pull up call records and attempt to determine what went wrong. It can talk to the local carriers who were part of the call and reconstruct an exact circuit-by-circuit map of how that call was placed, and each carrier can test that circuit for problems. You can be fairly well assured that all participants are going to try to fix the problem so that it doesn't happen again. That's part of what you are paying for.

Calls made over the Internet have no such assurances of quality or reliability. Your provider of Internet access services can only assure you a quality connection into the Internet; after your data crosses out of your ISP's equipment, your data packets are on their own. (We hope you have tough data packets — it's a rough world out there.)

In the not-so-distant future, you can expect to see companies offering some sort of industrial-strength Internet service that parallels the Internet but has more of the quality and reliability controls of the PSTN. You're probably going to have to pay more for it, too.

Telephone and videoconferencing are both very forgiving applications; if you lose a small piece of your data stream, you may merely clip a word or not refresh a portion of the video picture for a moment — not nearly as devastating as losing part of a spreadsheet or word processing file.

Privacy

Is someone listening? Probably not, but possibly. We're not paranoid, but the fact is that any data you send over the Internet travels an unpredictable and convoluted path through many switches before it reaches its final destination. On a theoretical level, somewhere along the way, in any of these switches, people can, if they really want to, intercept some of your data and read, listen to, or watch it.

Should you worry? Not really. We argue that privacy is very much a nonissue. In practice, all that anyone may obtain is snippets of information that don't make much sense without the bigger picture (pardon the pun).

Some Internet telephony products have options, such as encryption, to protect your privacy. See the discussion of PGPfone in Chapter 7 for more on this topic.

Cost: How long will it be free?

Strictly speaking, the Internet is not free. Someone has to pay for all those miles of fiber optic cable and wire and the servers and routers that make up the 'Net. At first, the government paid (sorry, a little history slipped in), but now the Internet is privately maintained and built. Your Internet Service Provider pays for its little part of it, and you, of course, pay the ISP.

But unlike the PSTN, the charge is not based upon the distance you send your data or what time of day you send it. In fact, with the advent of unlimited Internet access service pricing, you don't even pay for how long you use it or, up to a limit, how much data you send.

Some local telephone company access-line options, as we describe in Chapter 3, do carry per minute or per hour fees, however, so you could be hit with some add-on fees.

Overall, although the Internet is not really free, it sure is cheap. Now, how long can this low-cost status continue? No one really knows for sure. Many pundits, including us, believe that alternative Internets may offer similar connectivity options, with higher quality, reliability, customer services, and so on — for a slightly higher subscription fee.

Free (?) speech

Freedom of speech is surely a slippery issue when you think of the Internet these days. As we write, the Internet is on trial (to quote the headlines). Certain groups (some of whom know about the 'Net, some of whom really don't quite get it) want to regulate the Internet as they would a broadcast medium (television or radio, for example). Others see these rules as the first step on a slippery slope that infringes on the free speech of all 'Net users. So a battle rages.

We don't want to have to choose sides here, but we can say this: Regulating the 'Net is impractical at best and perhaps even impossible. You can log on from your computer in Arizona and, within a minute, be reading files from a computer in London or talking to someone in Kuala Lumpur, Malaysia. Laws that apply in Singapore, for example, don't apply in the United States or France. The global nature of the 'Net makes it almost immune to efforts of a single country to control it.

Freedom of speech is likely to become a big issue with Internet video especially, because a whole industry is ready to offer nude shows via the Internet for anyone with a credit card in hand.

Although a regulator may be able to control static text or graphics files at a Web site, dealing with real-time voice and video communications is a different story. No record is left behind, so unless the regulator can monitor everything as it happens, enforcement is impossible.

Note: Consider that you can pretty much say anything, about any subject, on the phone with impunity. If you use the phone to harass someone or to plan something illegal, you may be breaking the law — but no one tries to censor the phone system itself.

How Does an Internet Phone Call Work?

Making an Internet telephony or video call is a pretty sophisticated and complicated procedure, but the basic concepts aren't too difficult to understand. Say that you want to call your friend. Here's the 60-second version of what happens:

1. **You log onto the 'Net.**
2. **You start your software application.**
3. **You enter the name, e-mail address, or IP address of your friend.**

   Some programs enable you to select your friend's name from a list of options.
4. **You click on Enter.**

   The program connects you to your friend's computer, either directly or through an intermediate server.

From this point, the process continues as follows:

1. **Your computer takes your audio and/or video input from a microphone and camera and then digitizes it.**

   (*Digitize* means to convert the input from an analog to a digital signal.)
2. **The phone or video application samples and compresses the signal to make it smaller.**
3. **The program also converts the digital signal into packets that can be distributed across the Internet.**
4. **Your computer sends the packets to your friend's computer.**
5. **Your friend's computer reassembles and decompresses the packets.**
6. **Your friend's computer converts the digital signals back to analog and displays them on-screen or plays them through speakers.**

Connecting across the Internet

The connection process is a little more complicated than we make it sound.

Internet phone calls (and video calls, too) are controlled in one of two basic ways:

- Through a server
- Directly

Each method has advantages and disadvantages. Some programs use one or the other method; some programs offer you the choice of using either. (We discuss which techniques individual programs use in Chapters 6 and 7.)

Server connections

If you've seen any old movie or just about any *Green Acres* episode, you're no doubt familiar with the role an operator played in the early days of communications. To place a call, you rang up the operator, and she (in those days, it was almost always a she) used bulky telephone board cables to connect you with whomever you wanted to call.

A server connection operates in the same manner, but with an automated operator (a server) overseeing the process. Here's the order in which it works:

1. **Phone and video programs that use a server connection connect your computer to a central computer (the server) along with all the other users of the program who are currently online.**
2. **Your name appears on a common list of users that is viewed by everyone who is logged on.**
3. **You select the name of the person you want to call and click on the call button.**
4. **The server figures out where you both are located, in terms of IP addresses, and connects you.**

Some programs — for example, Internet Phone — use a server to establish the connection of your call but then route the actual call data (your voice) directly from person to person (or computer to computer, if you prefer) without having it pass through the server.

This method has the following advantages:

- It's easy.
- You don't need to know the other party's IP address, just that person's name.
- You can find lots of people who just want to chat — much like the text-based chat many 'Net users are already familiar with.

Server-based options have disadvantages, too, the following being among them:

- Servers can become so overloaded that you can't get on.
- The server can be down for repair, and you're unable to make calls.

- You may get bombarded with unwanted calls.
- Logging onto a server, searching through hundreds of names, and finding the person you want to talk to takes more time than calling someone with a program that uses direct route connections.

The problems with the server-based options are true issues in day-to-day calling and cannot be understated. With the proliferation of free, limited-use programs for people on the Internet to try, many people are making short, repeated requests to the servers, and this demand can bring any computer to its knees.

Direct connections

Direct connecting programs are similar to PSTN phone calls. You enter into a call window some sort of an address for the person you are calling (usually an IP address), and you are connected. The program takes care of navigating across the 'Net and connecting you with the person you want to talk with. No central computer is used to route or carry your calls across the 'Net.

Note: Earlier in this chapter, we discuss the "connectionless" nature of communicating over the Internet. Now we're talking about direct connection. We're not trying to confuse you — honest. In this case, direct connection means that the call data is not routed to a central point (a server) and then forwarded, but rather the call is connected from machine to machine across the 'Net. This direct-connect data still follows the same haphazard — and connectionless — path across the Internet.

The direct connection method has the following advantages:

- It's quick.
- Because it uses no central server, you avoid all the disadvantages of server connections.

And, of course, direct connections also have some disadvantages, as follows:

- IP addresses are hard to remember and can change.
- Chatting with random strangers is harder because you have no central meeting place.

Combining the best connection features: Directory services

Most Internet telephone and video products are beginning to move away from the pure server connection method (which we discuss earlier in this chapter) — in which both parties make a connection to a central point and then connect their calls through this server — in favor of directory services. Directory services go by different names depending upon the program you are using (for example, the Microsoft directory service is called ILS, or Internet Locator Service, and Netscape's is called DLS, or Directory Locator Service), but they all have the same basic function: They provide a constantly updated database of who's connected to the Internet and using a specific Internet telephony program.

When you log onto the 'Net and turn on your Internet phone, it sends a message to the directory service to supply some basic information about you (basically your name, your current IP address, and perhaps your e-mail address or other information you choose to provide), for updating the database. When you want to place a call (or when someone wants to call you), you typically have two choices:

- You can view a listing of people who've logged into the directory service, find the person you wish to call, and click on their name with your mouse. This is often done with your Web browser (for example, Netscape Navigator or Internet Explorer) and appears to work just like the server connection method. The big difference is that the call is actually made by the direct connection method — all the server does is figure out the other person's IP address for you.
- If you know exactly who you want to call, you can just enter the person's name (or in some cases the e-mail address) in a dialog box in your program. You never have to search through the directory listing or know the other party's IP address — the program does this for you!

It's not just Internet telephony vendors who are getting into the directory service business nowadays. A growing number of third-party companies offer Web-based "white page" services to users of Internet phones. Most of these companies, like Four11, started off as online e-mail directories and have branched into telephony. For the most part, these services are free and a good way to find people to call. As the number of users increases, we think you'll see even more of these white page services pop up on the Internet, with even more features and more ways of finding people you want to talk with.

Getting a Handle on Internet Conferencing Standards

One of the biggest changes that has occurred in the world of Internet telephony since we wrote the first edition of *Internet Telephony For Dummies* — and many changes have taken place — has been the establishment and adoption of standards. Back in the wild west early days of Internet telephony and videoconferencing, almost every program used its own proprietary method of sending data over the 'Net. If you had program A and your friend across the country had program B, you couldn't talk. Period, the end.

That wasn't a very good way to make Internet telephony a useful and widespread tool, but in the beginning, many of the companies that produced these products were hoping that *their* proprietary way of doing things would grow so popular that it would become the *de facto* standard — overwhelming success in the marketplace would lead to one program being "the one to have." Not surprisingly, this didn't happen.

So what did happen instead? Well, two of the biggest companies in the computer and Internet world (you may have heard of them — they're called Microsoft and Intel) got together and pushed a set of standards for multimedia Internet communications through all of the appropriate standards bodies — groups such as the International Telecommunications Union (ITU) and the Internet Engineering Task Force (IETF). So now we have an official standard in place — known as H.323 — that should make it easier to talk to other people over the 'Net. Pretty cool, huh?

Of course, H.323 is just one piece in the standards puzzle — the biggest, most important piece, but still just a piece. In the following sections, we tell you about H.323 and several related standards.

Why worry about standards?

Okay, you know that a standard exists. Should you worry about it? Or is it a done deal?

The answer to that question is an ambivalent yes — and no:

- **Yes.** The H.323 standard is still new. Not every program adheres to it, yet. In fact, most don't. So if you don't choose carefully, you could pick a program that can't yet talk to another one (which doesn't mean the program you choose won't be useful, but it may limit whom you can talk to). And even the programs that use the H.323 standard may not yet have implemented the entire standard, or they may do some things slightly differently — which results in incompatibility with other programs. To avoid surprises, it pays to do a little research and read about a program before you start using it.

- **No.** Many companies are moving quickly to adopt and incorporate H.323 into their products. Not every program is 100 percent in line with the standards yet, but as time passes (and time moves very quickly on the 'Net), standards incorporation and product interoperability will increase. We believe that before too long, you'll be able to pick and choose the Internet telephony program you use based on whatever criteria you like (price, ease of use, pretty interface, and so on) and not have to think twice about whether it will work with the product you didn't pick.

And the standards are . . .

Well, now is as good a time as any to give you a formal introduction to the standards we mention in the preceding sections. Don't worry, we won't drag this out on you. Just a brief hello to each of them, and you can be on your way.

H.323

H.323 is the mother of all Internet telephony standards, as you have no doubt divined by now. H.323 is an ITU-approved (remember? that's International Telecommunications Union — a group of sages in Geneva who approve proposed standards that make telecommunications stuff work together) standard that governs multimedia communications over packet-based networks (all sorts of packet-based networks, but primarily the Internet and corporate intranets). If you want to know the history, H.323 is part of a family of standards including H.320, which governs the same things over ISDN networks, and H.324, which rules over the POTS (Plain Old Telephone Service — basically regular phone lines) multimedia conferencing.

You could probably fill a whole book with arcane details of the H.323 standard — but we're pretty sure that wouldn't be a *...For Dummies* book. If you want to dig up more dirt on the standard, however, don't let us hold you back. If we were doing so, we'd check out the ITU's Web page at `http://www.itu.ch`, or maybe the IETF's page at `http://www.ietf.org`, or yet another group's page (the International Multimedia Telecommunications Consortium, or IMTC) at `http://www.imtc.org`. You can do the same if you'd like. Or you can just remember the name (we'll repeat it: H.323), and go on and find out how to use this neat stuff without sweating the details. That's what we'd do.

T.120

Among its many parts, H.323 incorporates another ITU standard called T.120. This standard governs multipoint data conferencing and collaboration. What the heck is that? Well, basically it's several people using computers to share data and information in a group effort (that's the collaboration part). For example, you may want to look at some drawings or documents with a colleague in another office. You could send faxes or ship hard copies of the documents via an overnight service. Or you could use your computers to view, comment upon, and edit the same document on your screen. Find an error, make a correction, and save an updated copy. Sure sounds a lot easier than a dozen faxes flying back and forth!

How important this standard is to you depends on why you're using an Internet telephony product in the first place. Calling grandma? Well, then you're probably not going to worry about T.120. Using Internet telephony in your workplace? In that case, you may very well think that T.120 is the greatest thing since sliced bread.

Codec standards

Codec? What the heck is a codec? A codec is a compression (the CO part) and decompression (the DEC part) algorithm. As we mention briefly in a previous section of this chapter, *compression* is the process by which your software squeezes down the size of your digitized voice or video picture so that it can fit efficiently over the sometimes crowded networks that make up the 'Net. The H.323 standard includes a couple of mandatory codecs that a product must include to adhere to the standard — specifically, an audio codec called G.711 and a video codec known as H.261.

The neat part about H.323, however, is that it allows developers to plug a bunch of other codecs into a program and let you (or the program itself) decide which ones to use. This is a good thing too, because the mandatory codecs we just mentioned are designed for use over high-speed network connections — neither of them is suitable for dial-up modem connections to the 'Net. The most commonly used codecs for modem users are the G.723.1 audio codec and the H.263 video codec. If you dig around inside the settings dialog boxes of your favorite program, however, you'll probably see that a whole bunch of codecs are available to you. In fact, the codec-design business is really starting to heat up, and all sorts of companies out there have scientists and engineers hard at work developing algorithms that can squeeze your audio and video down even more while giving you an even higher level of quality.

Directory standards

In the preceding sections, we tell you that standards are in place that generally govern how Internet telephony products talk to each other (H.323), how they enable you to collaborate (T.120), and how they encode your voice in video (the various codec standards). Only one piece of the puzzle is missing — how to find someone else to communicate with.

As we discuss earlier in this chapter, a directory service is the most common means for hooking two people together with Internet telephony software. Unfortunately, most programs use their own proprietary directory service system — so it's hard to use one program to find out the IP address of someone using a different program, even though the two programs are otherwise able to talk to each other.

However, we can see a light at the end of this tunnel. *Lightweight Directory Access Protocol,* or LDAP, is fast becoming the standard way for Internet telephone products to access directory servers. A new version of this standard, LDAP 3.0, is currently pushing its way through the standards bodies and should be implemented by the summer of 1997. Until then, a bit of confusion will remain in the directory service world.

You can get around any directory service incompatibilities by using one of the third-party directory services, such as Four11 (`http://www.four11.com`). These systems use some techno-wizardry of their own to connect phones that might otherwise have a hard time finding each other.

In case you were wondering, the *Lightweight* part of the name LDAP has to do with the complexity of the standard. LDAP is based on a sophisticated directory standard known as X.500. While developing LDAP, engineers took a lot of features and other stuff that wasn't really necessary out of X.500. In a sense, LDAP is X.500 on a diet — a real lightweight compared to the rather hefty X.500.

Gateways to the World

Besides the H.323 standard (which may or may not excite you, but we find it pretty neat), the biggest development in Internet telephony since the first edition of this book has been the development of gateways. So, what's a gateway, you may wonder? In general terms, a gateway is a device that connects two different kinds of networks and performs all the translations necessary to make communications work. For example, if you have an office full of Macintosh computers, they may be networked together using AppleTalk as the network protocol. If you want to connect this network to the Internet (which, as you know, uses TCP/IP as its protocol), you could use a gateway. Plug the Internet connection to one side of the gateway, and the AppleTalk network to the other side and — voilà! — you're connected.

In Internet telephony, gateways perform the same sort of function. You can find different gateways that do different things, but in general they all have this trait in common: On one side of the gateway is a TCP/IP network (the Internet, or a corporate intranet, for example), and on the other side is a different network.

Gateways already out in the marketplace, or soon to be, include

- Gateways that connect calls made over the Internet from your computer to the regular phone system — so you can call grandma from your PC even if she doesn't have one
- Gateways that connect the telephone systems in offices to a company's intranet, enabling users to call or fax between offices for free
- Gateways that connect PCs on the Internet to stand-alone conferencing systems (like T.120-compatible data conferencing systems or compatible videoconferencing systems)
- Gateways that are installed in the local offices of a telephone company, or one of its competitors, allowing you to make phone-to-phone long-distance calls over the 'Net

With the development of these gateway products, the line between the Internet and the traditional telephone network is becoming increasingly blurred. In fact, many telephone companies themselves are beginning to show an interest in Internet telephony and other variations on this theme. MCI, for example, has recently announced a new group of services known collectively as VAULT. The aim of VAULT is to combine MCI's traditional long-distance network and its Internet network (MCI is one of the seven or so companies that provide most of the "backbone" of the Internet) in a seamless and transparent way — so you could be talking with an MCI VAULT customer from your phone or PC and not even know whether the customer was on the regular telephone network, the Internet, or both (nor would you care!).

All in all, gateways are some pretty neat stuff. We talk about them in more detail in Part III of the book.

Extra Credit: Fax over the 'Net

Another new and exciting development in Internet telephony has occurred during the last year or so — the development of hardware and software that lets you send faxes over the 'Net. This development should be no surprise when you actually think about it:

- Sending a fax, in a lot of ways, isn't so different from making a phone call (something that we all know is increasingly moving to the Internet).

- Faxes are basically data transmissions — and what's better for sending data than the Internet?
- Faxes really aren't as affected by the bandwidth or delay issues that sometimes make voice or video calls over the 'Net difficult — as long as it all gets there, who cares if it takes a few extra seconds?
- Faxes cost businesses a lot of money (upwards of 40 percent of their total telephone costs), so companies have lots of incentive to develop and market products that will save these businesses some money.

In Part III of this book, we talk more about how to fax over the 'Net, but here's the quick rundown on how it's generally done:

- **E-mail to fax machines:** This is the "old-fashioned" way of faxing over the 'Net, typically seen as an extra-charge service provided by online services. Basically you would send the material you wanted to fax inside an e-mail message to a specific address, and on the far end, it would be spit out and faxed to the fax machine you specified, over the PSTN. Some new services have recently surfaced, however, that turn this equation on its head — they let you have all your incoming faxes sent to one number, and then the faxes are forwarded to you (wherever you may be) via e-mail.
- **Faxing from the desktop to fax machines:** At first glance, this may sound just like the e-mail to fax machine method, but it differs in one way. In this scenario, you have special client software on your desktop (or laptop, for that matter) computer that enables you to fax a document from just about any application through a gateway and over the Internet or your company's intranet. Instead of printing the document and walking over to the fax machine, you basically "print" the document as an Internet fax. The gateway that allows you to do this can be located in your corporate network or operated by a service provider that you access via your modem connection.
- **Fax services (fax machine to fax machine):** This is the one that makes the folks down in accounting who pay your company's telecommunications bills very happy. Many of the gateway products we discuss in the previous section accept calls from fax machines as well as regular telephones. So those hundreds of millions of fax machines sitting out there don't have to rely solely on the PSTN anymore — they can plug into the gateway and send faxes on the cheap to other fax machines connected to a gateway on the far end of your call. Pretty neat stuff and potentially a huge money-saver.

Chapter 3

Getting Yourself on the Internet

In This Chapter

- Plugging into the 'Net
- Picking an Internet Service Provider (ISP)
- Making sure that your ISP supports telephony and video
- Figuring out your phone line requirements

After you know a little about the Internet and a little more about Internet telephony and video, you are probably ready and eager to get online and find out what all the fuss is about. Your first, logical step is to get yourself on the Internet.

Getting onto the Internet

If you're already online, at least browse through this chapter to see whether you're missing anything. (For example, if you're using a modem, have you considered an *ISDN* — Integrated Services Digital Network — line? You read more about them later in this chapter.) If you're a true novice, read on and get some advice on plugging in.

If you're not that familiar with other aspects of using the Internet (such as the World Wide Web, FTPing, or e-mail), we point you in the direction of a number of fine *...For Dummies* books, such as *The Internet For Dummies,* 4th Edition, by John Levine, Carol Baroudi, and Margaret Levine Young; or *The Internet For Macs For Dummies,* 2nd Edition, by Charles Seiter (both published by IDG Books Worldwide, Inc.). In these books, you get lots of juicy details about all the other things you can do on the 'Net to complement your telephony and video programs.

Figuring out who the providers are

In Chapter 2, we hint about the increasing number of Internet access services, including independent Internet firms, local telephone companies, long-distance companies, cable companies, and the well-known online service providers. Historically, the term *ISP* (Internet Service Provider) referred to the independent businesses with names such as Netcom and Earthlink. However, everyone who is anyone in the communications universe wants to sell you Internet access service and offers you the convenience of bundled services (by combining local, long-distance, cellular, and Internet access). Consequently, when we use the term ISP in this book, we use it in the most generic sense.

The traditional Internet access providers

Traditionally, ISPs came in two basic flavors: pure ISPs, such as Earthlink and Netcom Online Services, who plugged you directly into the Internet and enabled you to find your own favorite spots; and content providers, such as America Online (AOL) and CompuServe, who had their own proprietary networks and content. As time went by, the line between these two groups became increasingly blurred.

Not so long ago — say, in 1994 — users of AOL, CompuServe, Prodigy, and other online services were stuck within the proprietary networks of their providers. Sure, you could send e-mail across the Internet and maybe use a few other Internet tools, such as FTP or Gopher, but for the most part, you were cloistered from the vast jungle of the Internet.

The explosion of the World Wide Web changed all that. You couldn't go a day in the last two years without hearing something about the Internet on the news, reading about it in *The Wall Street Journal,* or even seeing it in advertisements during *Seinfeld.* People were clamoring for access to the Internet, and the proprietary services weren't about to miss out on a money-making opportunity.

As hundreds of new companies, big and small, started offering dial-up Internet access, the mainline proprietary services added Internet gateways to their old, and quite popular, online interfaces. These gateways enable you to do almost everything that someone with true direct access to the Internet can do, such as using Netscape Navigator to browse the Web.

The bottom line is that you can get your Internet connection from a variety of sources now — from IBM to CompuServe to some small company in your neighborhood.

The new(ish) kids on the block

The growth of the Internet and the unbelievable demand for access by folks all around the world have made many companies that were previously on the periphery stand up and take notice. The traditional model of an independent ISP with a sole function in life of providing 'Net access has gone by the wayside, while long-distance and local telephone carriers and cable TV companies are all rolling out their own Internet access offerings.

These companies all want to be your one-stop-shopping communications provider. They intend to offer you both the local access line (now the province of the local telcos) and Internet access (the ISP's piece of the action), as well as local and long-distance telephone service, television service, and even the hardware you need to get set up.

Is this a good thing? Well, for some smaller ISPs, and even for some big ones, this rush of competition could be the death knell, but for you, the consumer, it's great news. Prices will continue to drop, and high-speed access options will increase as these giants fight for your business. We can hardly wait.

Getting ISP support for telephony and video

As Nicholas Negroponte, head of the world-famous MIT Media Lab, said in his book, *Being Digital* (which you really should read — even though IDG Books Worldwide didn't put it out!), "bits are bits."

The same is true for Internet telephony and video. The bits of information flowing into and out of your computer are pretty much all the same, as far as the 'Net is concerned. It doesn't matter if you are making a phone call, watching an on-demand video program, transferring files, or sending e-mail.

Your ISP transfers these bits across the 'Net, from your computer to their destination according to the protocol that the application picks (TCP or UDP, as we discuss in Chapter 2). The 'Net is dumb in that it doesn't really care about the content of those bits. It just finds out where the bits are going and sends them there. Of course, the reality of how bits find their way from point A to point B is a little more complicated, but the process is transparent to the user.

Picking an ISP

Still, some providers can be more Internet-telephony-friendly than others. Many factors can influence how friendly your ISP is to you as a user of Internet telephony and video programs.

In deciding which ISP you want to carry your Internet traffic, consider the following major buying criteria:

- PPP/SLIP connections
- Access speed
- Access points
- Technical support
- Price

If you have access to the 'Net from work or a friend's house, check out the excellent piece on finding the best Internet Access Provider on c|net (a great online resource, by the way) at `http://www.cnet.com/Content/Features/Dlife/Iap/index.html`.

As companies such as AT&T and MCI begin to enter the Internet access market (and they are doing so in a big way), you may also want to consider what kind of bundled packages (long-distance and 'Net access all in a one-price package, for example) and bonus award programs they offer. Expect the local and long-distance companies to be very aggressive in courting your Internet access business.

As long as your ISP and your specific account meet the guidelines we suggest in this book, such as supporting a Serial Line Interface Potocol (SLIP) or Point-to-Point Protocol (PPP) access and providing adequate bandwidth for your connection (read about modems and access in a few pages), you can use 'Net telephony and video applications. We review our buying criteria in the next few sections.

SLIP versus PPP

Serial Line Interface Protocol (SLIP) and Point-to-Point protocol (PPP) are two methods for accessing the 'Net by dial-up connections. When you use a SLIP or a PPP connection, your computer is actually on the Internet, with an IP address assigned to your computer. Older connection methods, like shell accounts, connect your computer with one that is on the Internet, but they don't really provide you with the access you need for utilizing Internet telephony and video programs.

Make sure that you sign up for a full Internet access account with either SLIP or PPP in order for your Internet telephony or video program to work.

Access speeds

When we wrote the first edition of this book — less than a year ago — we warned you that you should be sure your ISP offered 28.8 Kbps modem connections. Well, that was then — now you should expect support for a minimum of that speed for dial-up accounts, and many ISPs support 33.6 Kbps modems. In fact, the big question for dial-up accounts now is whether they will support the forthcoming 56 Kbps modems. (We talk more about 56 Kbps modems in Chapter 4.)

Having a 28.8 or 33.6 Kbps modem and paying for 28.8/33.6 Kbps SLIP or PPP service doesn't mean that you are always getting a connection this fast. As you see when we cover 'Net telephone and video programs in more detail, the actual bandwidth of your connection greatly influences the quality of your phone and video calls. Factors such as heavy traffic, poor phone line connections, and inadequate ISP facilities can limit your connection speed and, subsequently, reduce your ability to optimize your program's connections. The world is full of millions of very long, very old, poor-quality local lines. You may want to keep an eye on your modem's actual throughput to determine where any problems may lie.

We like external modems with displays that tell you your connect and subsequent ongoing throughput speeds. The lights tell you whether poor-quality lines are causing the signal to drop back to 19.2 Kbps or 14.4 Kbps speeds. You can also tell whether you logged onto a wrong port at your ISP if anything but 28.8 Kbps appears on log on.

Port availability

The number of modem ports available in your area is also a big concern. Many ISPs overstate their capabilities. Do they have enough lines to handle the calls? How about high-speed modems? America Online's well-publicized woes in early 1997 are but one example of this oversubscription problem.

Many ISPs do not share their operating secrets, but you can check with your friends and local discussion groups about this topic. Internet users never face a shortage of commentary on how easily they can get onto various ISP platforms.

Access points: local, long distance, and toll free

Note: If you are looking into using an Internet telephone program to save yourself money on long-distance calls, you'd better make sure that you are not paying more for your Internet access than you would on long-distance charges.

In addition to paying the ISP for the service (which we discuss in a second), you must pay the phone company something for the link between your home and the ISP's *point of presence,* or *POP.* The usual link choices are *local, long distance,* and *800 (or 888) number.*

- **Local:** If your provider's nearest POP is within your local calling area, you shouldn't incur an additional per-minute charge.

  Make sure that you are indeed in the local calling area by checking the front of your phone book or calling the operator and asking. Per-minute rates in some areas are about a dime a minute, so your phone bill could add up fast!

- **Long-Distance:** Your POP is outside of your local calling area. You ring up per-minute tolls on your telephone bill. You probably want to check on other options.

- **800 Access:** Some providers offer access via an 800 or the new 888 toll-free numbers. But don't be surprised to find out that many of these numbers are not free; your provider charges you an extra per-minute fee along with your monthly subscription charge. And the fee may be higher than the toll charges you'd pay if you just dialed a nonlocal access number directly.

Obviously, local access is the preferred, and most frequently chosen, option, but if you live in a rural area, you may be forced by geography to use another method.

Technical support

Ever buy a new program, have a problem with it, and call the developer's technical support hotline? (Admitting it is okay; we've done it, too.) How many buttons did you have to press to get through to a real person? How long were you on hold? How much hair did you pull out? How many hairs turned gray while you were waiting?

We don't have to tell you how important good technical support can be. Make sure that the provider you choose has 24-hour-a-day, seven-day-a-week support by telephone. Unfortunately, the phenomenal growth of many ISPs has left them having too many customers and too few support people.

Note: Support by e-mail or online help documents can be great, but if you can't even get online, how are they going to do you any good?

Again, word of mouth is the best way to find out about who's on top with technical support.

The bottom line: prices

Like most everything else in the world of computers, prices for 'Net access have been steadily plummeting. In 1994 or 1995, for example, an average dial-up account with local access may have cost you $50 or $60 a month with only 20 or so free hours. As the Web has hit the mass market, however, competition has clobbered prices. Since AT&T announced its unlimited access for $19.95 a month last year, just about every major Internet Service Provider has matched or bettered this price.

As the number of competitors grows, not only do prices drop, but the number of local POPs increases. See, your economics professor was right — competition is good!

Choices, choices, choices

How do you find the best ISP for you?

Can the prices stay this low?

Unlimited Internet Access for 20 bucks a month. That's the promise that AT&T made with its WorldNet service last year. Consumers, by the hundreds of thousands, took them up on the offer. All the other major ISPs then dropped their prices to match.

Sounds great, huh? Well for us, the end users, it is. Unfortunately, for the ISP's networks, unlimited access has been like dumping off a bus full of hungry teenage boys in front of an all-you-can-eat joint. People dialed in and stayed on for days at a time. Networks got congested, modem banks got full, and people couldn't get through. This congestion forced ISPs to make a decision: Buy more modems and network capacity or have a bunch of angry customers facing busy signals every time they tried to connect. Increasing capacity seems like the proper response — but the low flat rates mean that revenues don't increase nearly as fast as the need to increase capacity.

A few brave ISPs are starting to increase their monthly fees to reflect their increasing costs — while promising better service as a benefit for paying a bit more each month. Others have abandoned the consumer marketplace and positioned their service offerings in the higher-profit business market. So are prices going to stay low, or will they creep back up?

We're not crystal-ball gazers, but if we had to guess we'd say this: Some of the major ISPs will continue to offer the $19.95 unlimited access model, using it as a loss leader to get your business for their other services. Others will raise prices while offering higher quality and additional services as their means of attracting your business. Still others will go broke and either go out of business or be snapped up by competitors.

Bottom Line: Prices will remain low, but your choice of ISPs will probably dwindle from thousands to just a handful.

- **Ask your friends.** If all your friends are complaining about their ISPs, choose the one that gets the fewest complaints rather than the most praise.
- **Call the big providers.**
- **Call local providers.** Scour through your local paper or the free computer magazines (the kind you find piled up by the door of your local coffeehouse).
- **Go to The List on the Web at** `http://www.thelist.com`**.** This list encompasses more than a thousand ISPs and is searchable by country, state, and area code. (See Figure 3-1.)

You can find our listing of the major ISPs in Appendix A.

Figure 3-1: The List.

Gauging Your Telephone Line Requirements

You thought that Internet telephony enabled you to bypass the telephone system, didn't you? Well, it does, to a degree, but you do need a telephone line. You have two options:

- An analog phone line with a modem (the best bet for most people because it's cheap)
- A digital line (ISDN), which costs a bit more but pays you back with increased speed and flexibility

You could get a high-speed, dedicated, private-line connection, but that can be a bit pricey. The choice is yours — read on, and we help you determine which is best for you.

Being analog

Sure, the 'Net is all about the digital future, but the telephone lines running into most of our homes are analog.

To make these analog lines transmit and receive the digital signals that your computer uses, you need a modem (MOdulator and DEModulator, literally) to act as an interface device. (We look at modems in more detail in Chapter 4.)

We don't have much to say about buying a normal telephone line. In fact, if you don't mind tying up your existing telephone line, use that one. The biggest problem that most people have with this is dealing with call waiting.

If you have call waiting on your telephone line and are using a modem at the time, the tone notifying you that a call is on hold can mess up your transmission and probably even disconnect you.

You can easily disable call waiting for the duration of the call. Go to the dialing setup portion of the communications program that you use to dial into your ISP. (In Windows 95, look in the Dial-Up Networking application; on a Mac, check in the PPP or FreePPP control panel.) Enter ***70** before the local access number if you have touch-tone service or **1170** if you have pulse service.

Consider getting a second phone line for your home that you can dedicate to your modem. For 10 or 15 bucks a month, you can avoid call-waiting hassles and still get that call from your friend who hasn't yet discovered Internet telephony — while you're online.

Deciding on digital (I want my ISDN now)

The pair of wires connecting your house or office to the outside world can be digitized for higher-quality service. With a digital line, all data is converted into a series of 1s and 0s, which are easier to regenerate and, therefore, maintain the signal over long distances. The analog wave, on the other hand, is tough to regenerate accurately and degrades over the long haul.

Although many phone companies are promoting a service known as Integrated Services Digital Network (ISDN) as the next big thing, the technology has actually been around for quite some time but has never really taken off (mostly because of marketing, deployment, and cost issues). The local telephone companies, facing future competition from competitors such as cable and wireless companies, are finally giving ISDN the organized, big marketing push it deserves.

ISDN offers the following advantages:

- **More bandwidth:** You get almost five times more bandwidth on ISDN than you do on the fastest analog modems (though the forthcoming 56K modems may decrease this deficit).
- **More accessible:** ISDN is now widely available in many regions of the U.S.

- **More economical:** Users are not having to pay backhaul charges or central office equipment expenses to get ISDN in remote areas. Also, new tariffs are offering unlimited ISDN usage options, although in some parts of the country this can approach $200 per month.
- **More capable:** ISDN is especially good for Internet telephony and video applications because the higher levels of bandwidth (which require less intrusive compression methods) make voices sound crisp and the video appear more lifelike.

Getting and paying for digital service

Your local phone company is probably glad to sell you an ISDN line, as most phone exchanges (your local switching office) now have ISDN availability. Some telcos, such as Pacific Bell, have ISDN available to more than 90 percent of their customers, while others are down in the 70 to 80 percent range. If you live on top of a mountain 20 miles from the nearest town, you may be out of luck, but if you live in a slightly more populated region you should be okay.

Note: Calling your local phone company for information about ISDN can be an interesting experience, at best. Many of the customer service personnel are not quite up to speed yet on what ISDN is all about, especially at some of the smaller companies. We actually experienced this phenomenon when a phone call to our carrier inquiring about ISDN was returned to a "Mr. Isden." Happily, many phone companies have set up special ISDN hotlines. We list these in the following table.

Carrier	Number
Ameritech	800-832-6328
Bell Atlantic	800-570-4736
Bell South	800-428-4736
GTE	800-448-3795
NYNEX	800-483-4926
Pacific Bell	800-472-4736
Southwestern Bell	800-792-4736
US West	800-898-9675

Note: Many of the phone companies have ISDN availability information on their Web pages — to find out whether you can get ISDN, you just have to type your home phone number into a form and click on a button. You may notice that many of the phone numbers listed in the preceding table end with 4736 — this spells out ISDN on your phone keypad.

Really simple ISDN basics

ISDN, in its most basic form, is often referred to as BRI, or *B*asic (naturally) *R*ate *I*nterface. It uses, in most cases, the same copper wires that run into your house right now, but by using digital signals, rather than analog ones, BRI can give you up to 128 Kbps of speed.

A BRI connection actually consists of three channels. Two big ones, known as *B* (for bearer) channels carry the data, while the little one, the *D* (or data) channel, carries signaling information (that tells the phone company how to direct your call) or other kinds of data.

Each B channel can carry 56 or 64 Kbps of data. (The amount depends on your local phone company's switches.) The terminal adapter (TA) that hooks your computer into your ISDN line, much like a modem on a conventional phone line, can combine these two channels (by using bonding or multilinking, for you hardcore geeks) to give you the maximum data throughput, 128 Kbps.

One big advantage of having these multiple channels is that you can do more than one thing at a time over an ISDN line. For example, if you are using your TA to transmit data on both B channels and you receive a regular telephone call (not a 'Net phone call), the TA automatically reduces the data flow to one channel and enables you to take the phone call on the other channel. No more busy signals if your mom calls and you're online. You need to decide for yourself whether that's an advantage or not.

If you want to get real smart on ISDN real fast, read *ISDN For Dummies,* by David Angell (published by IDG Books Wordwide, Inc.), or if you're online already, check out Pacific Bell's excellent info site at `http://www.pacbell.com//Products/SDS-ISDN/Mag/index.html`.

The key to ISDN is its cost. ISDN used to be really expensive everywhere; it still may be expensive in your area. Monthly fees range from the reasonable ($25) to the ridiculous ($200), depending on where you are in the world. ISDN is priced in much the same way as your regular telephone service, with an installation charge, a monthly service fee, and possible usage rates. It is, however, charged at rates that are slightly to significantly higher, depending on your telco. In addition, per-minute charges (from a penny to nine cents) almost always apply (unlike the local flat rate services available in most places for an analog telephone line) unless you buy the unlimited service option.

Installation fees range from $40 to nearly $300, although many telcos offer discounts if you sign up for a long period of time (usually one or two years). Each telephone company probably has a range of options. A typical offering may be $30 per month, 10 hours free, two cents a minute thereafter; or $60 per month, 20 hours free, two cents a minute thereafter; or $120 unlimited usage.

Watch out for firewalls

If you have a direct connection to the 'Net, your computer may be behind a firewall. This is good news for your office's network administrator, because a *firewall* is a piece of hardware or software (or both) that controls access between the outside world (the 'Net) and the relative safety of home (the local area network, or *LAN*, on which your computer resides). A firewall enables the computer people in the office to get at least a little bit of sleep at night, as it reduces the number of nightmares they may have about some bad guy (or girl — but they are usually guys who need a shave and should probably quit drinking highly caffeinated cola) breaking into your network and wreaking havoc.

Firewalls are bad news, however, for users of Internet telephone and video programs because most firewalls block access to UDP ports. And almost all Internet phone and video programs use UDP as their transmission protocol because UDP enables a faster flow of data than TCP. The Internet telephony and video software companies have not really come up with a solid solution to this puzzle yet; after all, to a degree, the firewall is just doing its job — that is, keeping unwanted people out of the LAN.

So if you are directly connected to the Internet at a corporate site that uses a firewall, please talk to that harried-looking person in the back room (that is, the network administrator) before you lay out any cash getting set up for Internet telephony and video. Your network administrator may or may not be willing — or even able — to solve this problem for you.

Despite its higher initial costs, ISDN can be cost-effective, especially if you use it to replace two phone lines in your home office (no need for separate data, voice, and fax lines) or if the time that you save in downloads and 'Net access is important to you. But you have to weigh the costs heavily. Remember that you can get residential long-distance telephone service for about a dime a minute (at least according to Candice Bergen)!

Getting a direct connection

Although modems and ISDN are both excellent methods of connecting to the Internet, they can't compete with a dedicated, direct connection. A direct connection means that you have arranged with your ISP to have a point-to-point, digital line that hooks your computer (or your local area network if you are in a business) into the Internet. With a direct connection, your machine is always on the Internet and is, in fact, part of the Internet.

These direct-connection lines are usually very fast, often with speeds measured in the megabits per second (as opposed to kilobits per second of modem and ISDN connections). Of course, you must share this speed with other people on your local network, but you should still end up with much more bandwidth than any dial-up user can hope to have.

In addition to speed, a direct connection can give you a permanent address on the Internet. Dial-up users typically receive a different address every time they log on. As you see when we begin discussing the various phone programs, this variation can be a real pain if you're trying to find someone or if someone is trying to find you. Imagine your phone number changing every time you picked up the phone, and you can begin to understand the problem.

The bottom line is to get the fastest 'Net access you can afford and to shop around for an unlimited access account so that you don't rack up a great deal of per-minute charges. If you're already on the Internet, by all means try out some Internet telephony and video programs with what you're already using and upgrade later if necessary.

Chapter 4

Getting Outfitted for Internet Telephony

In This Chapter

- Figuring out your peripherals
- Sounding off about speakers and microphones
- Customizing connectivity options
- Processing the signal

Leading edge technology can be good and bad.

- **What's good?** Early adopting means that you can be on the cusp of something totally new and that you can be among the first adventurers to take advantage of tomorrow's technology.
- **What's bad?** You may have to be a real power user with in-depth knowledge of the hardware, software, and networking to get the new stuff to boot up. You probably also need to spend a good deal of your time, money, and daily frustration quotient just to keep it going.

Well, the great news is that Internet telephony isn't like that (for the most part):

- Many people already have the hardware and software required to run these programs sitting on their desks right now.
- Even if you don't own all the equipment, you probably can with a pretty small investment (barely a hundred bucks if you already have a modem).
- The software is usually very easy to set up — just download it from the vendor's Web site, and you are ready to go.

Better still, you can try almost all of the programs that we're about to mention for free. Some of them are entirely free, and some enable you to try out a demo version, which you can then pay to activate if you like it.

Don't expect to find the popular telephony programs in your local software store, because most of them are available only from the 'Net. How's that for futuristic?

This chapter can help you check out your hardware and software situation to make sure that you're properly equipped to get online and start downloading. You could be talking on the Internet today!

Setting Your Standards

You can find an Internet telephony program for almost any computing platform and a video program for most platforms. Regardless of your platform for Internet telephony, some general requirements apply for things like modems and peripherals.

Each element of your hardware and software is critical for the Internet telephony and video applications, so don't breeze through this chapter too quickly — the "garbage in, garbage out" (GIGO) principle very much applies to all aspects of Internet telephony. No matter how fancy the program and how immensely powerful the computer on the other end, if you don't have a decent system on your end, the recipient gets nothing but muffled audio and video.

When in Doubt, Go with the Best

In general, the Internet telephony and video programs are being designed for the leading edge platforms. In some cases, their software requirements are too sophisticated for two- or three-year-old machines; these applications require the most up-to-date hardware you can get. Other pieces of software are more forgiving and can deal with a fairly broad range of computing platforms. We discuss all of these and tell you what platforms are required for which systems.

The golden digital signal processing (DSP) rule

In Chapter 2, we talk a lot about sampling and compression — two techniques that help to cram a lot of information about your voice and video signals over the wee small pipes going through the Internet.

Parlez-vous GSM

GSM, or Global System for Mobile Communications, was created by a group of European telephone companies as a standard for their new generation of digital cellular phones. The standard was developed because cellular phones face the same problem as Internet phones — limited bandwidth. In the case of mobile communications, cellular phones get only a small part of the usable spectrum. The rest is divvied up for broadcast television, radio, and a host of other uses.

As you are no doubt aware, cellular phones have been experiencing the same kind of exponential growth as the 'Net. And this rapid growth has left users in populated areas stuck without enough bandwidth — their own little piece of the airwaves is just too crowded.

The solution, of course, should sound familiar to you — compression! GSM uses compression technology to squeeze more calls onto the air.

Turns out that the GSM algorithm is publicly available, so several creators of Internet phone programs have incorporated it into their programs.

However, sampling and compression require a lot of sheer processing power to work fast enough to avoid the delays that can ruin the natural flow of your conversations. This processing power, which is in your computer's internal central processing unit (CPU), can require a lot of work for the CPU, especially if you are using a higher quality technique. This condition is true for both Macs and PCs.

For example, the Global System for Mobile Communications (GSM) protocol (see the sidebar "*Parlez-vous* GSM"), which is used in Europe for encoding transmissions over digital cellular phones (and by a few of the Personal Communications System (PCS) companies starting up in the U.S. as well), is also used by several Internet telephone products. It's a high-quality, good-sounding technique, but it can place an intense load on your computer.

How do these cellular phones, which cost $300, do what it takes a $1,500 computer to do? They use a special computer chip, a *digital signal processing (DSP) chip,* which does nothing but encoding. In other words, the phones rely on specially designed hardware, rather than software, to do their work. Don't feel bad, though, because users can't play Doom on their cellular phones, no matter how much they pay for them.

John Walker, the founder of Autodesk software (the AutoCAD company), did some calculations when he was developing his Speak Freely Internet telephone program and figured out that you need a minimum of a 50 MHz 486 computer to keep up with the demands of GSM. If you try to use a GSM compression scheme on a slower computer, you simply run out of computer horsepower.

For maximum performance across all the Internet telephony and video products, get the fastest machine you can afford — if you are buying a new computer. MMX Pentium PCs and PowerPC Macintoshes and clones with the latest operating systems (meaning Windows 95 or Mac System 7.6) and a bunch of RAM make life much easier — as is true in most areas of personal computing. Buying the best "future proofs" your investment. Don't be swayed by inventory reduction sales on older merchandise — that stuff is on sale for a reason.

Here's a corollary to the GIGO principle: Slower machines don't work as well as faster machines. Slower machines cause you to use less powerful encoding techniques that are likely to "lose" more of your voice and send a lower-quality signal across the 'Net.

Before you put this book down in disgust, however, hang on for one more second. These fast machines make things work better, but you can still get into Internet telephony and video with a lesser machine. You don't need to throw out what you have and spend a fortune to get going! In fact, we made good-sounding, high-quality calls all over the world using a dial-up connection and a several-years-old 486 machine. The trick is setting up the machine correctly.

After all, if you're trying to save some money by using Internet telephony, you don't want to spend thousands of dollars on a new machine and burn through near-long-distance rates with an expensive ISDN line sporting per-minute rates. (More about ISDN later in this chapter.)

The best-sounding call we've experienced to date went from a Power Mac in New Jersey to another Power Mac in California; one machine was connected to the 'Net by an ISDN line, and the other by our office's dedicated Internet connection. The quality was high, the full duplex worked perfectly, and the conversation was smooth and natural.

Some programs require a significantly larger dollar investment than you may be prepared to make; we point these out along the way.

In the future, you can expect computers to have dedicated hardware devices for squeezing your voice or video signal to fit into the bandwidth of your Internet connection (compression) and then turning the signal into a series of data packets that can be transmitted across the Internet (encoding). For example, Intel's new chip technology, MMX, does just that. (See the sidebar "Multimedia speed for the masses.")

The gateways that we talk about in Part III, "Internet Telephony, Not Just for PCs Anymore," are a more extreme example of this trend toward using dedicated devices for Internet telephony — stand-alone computers that do nothing but take conventional phone and fax calls and convert them into Internet calls.

Multimedia speed for the masses

Intel, the ruler of the microprocessor world, has a new technology that makes Internet telephony and video faster and easier. The newest version of the Pentium processor, used by the vast majority of recent PCs, incorporates Intel's new MMX technology. MMX means that your computer's CPU has a built-in digital signal processing (DSP) chip to perform high-speed processing of audio and video signals without affecting the performance of the CPU's other functions.

TCP/IP network and node address

Although the 'Net is the big daddy of TCP/IP networks, you may want to try to use Internet telephony and video on other kinds of TCP/IP networks.

Note: Your *node address* is just a techie way of saying you need to have your own IP address. Not that IP address isn't techie enough, and because it's an important concept, you may want to review the information in Chapter 1.

For example, your company may have a TCP/IP local area network (LAN) or even a wide area network (WAN) across several locations. You can use your 'Net phone and video applications across these networks as well, so your boss can call you while you're trying to hide in the corner and perfect your Minesweeper game technique.

On the Internet, you need to have either a dedicated connection or a dial-up SLIP or PPP connection. (Go to the previous chapter if you don't feel warm and fuzzy with these terms.) This connection is something you ask your Internet Service Provider (ISP) for when you sign up.

Using a SLIP or PPP account with your ISP gives you an IP address. A simple shell account does not, nor do many online service accounts.

Remember that a dedicated Internet connection usually gives you a fixed IP address, while a dial-up account gives you a slightly different address every time to log on. Keeping this fact in mind is worthwhile because different software programs work around this situation in different ways. (More on this topic shortly; however, you may also want to turn to the server versus direct connection discussion in Chapter 2).

10MB hard disk space

Hard disk space is like chocolate; the box gets eaten up really fast. Before deciding to install an Internet telephone or videoconferencing program on your machine, make sure that you have at least 10MB available. The program itself takes up only 2 or 3MB of disk space, at most, but you need some extra breathing room for the installation. Of course, if you take our advice, you'll be downloading demos of several programs from the 'Net, and you'll need even more room. This one is a no-brainer, so use your good judgment here. The more the merrier.

Connectivity hardware

Your modem is the link between your computer and the world outside, so don't ignore its importance. In the previous chapter, we discuss the options you have for connecting yourself to your ISP. Modems are certainly the most popular and common means at the present time. In the coming year, you'll begin to see some new options that will make your modem, and even ISDN, seem like some sort of prehistoric contraption. (We discuss these topics in Chapter 24.)

Until these options come to your neighborhood, here's what to look for when you are ready to select a modem; remember, this time is not one to skimp.

Fast and faster

Most modems sold today run at the speed of 33.6 Kbps, although 28.8 Kbps models are still available. If you already have an older 14.4 Kbps modem and are planning on using only audio programs, you can get by if you have to. But if you are in the market for a new modem or are planning on using videoconferencing software, skip the few remaining 14.4 Kbps models and go right for at least 28.8 Kbps. Buying a slower modem is an exercise in false economy.

The amount of time you save in downloads and the amount of frustration you save while waiting for Web pages to load on your computer are more than worth the extra money. The added bandwidth of a faster modem enables your telephony program to use a lower compression of audio signals and usually offers a better sound quality. As for video, forget it if you don't have at least a 28.8 Kbps model because it simply won't work with any kind of quality — no matter what the brochures say.

If you can find a 33.6 Kbps modem, even better — just realize that you may not be able to make use of all this speed on your Internet access provider systems — many still max out at 28.8.

Just because your modem says 28.8 Kbps on the case, don't think that you always get a connection that fast. Modem connections are vulnerable to many variables, not the least of which is the condition of your local telephone network lines.

If you follow the computer industry news at all — and we're guessing that you probably do — you've no doubt heard of the new 56K modems that are just starting to come onto the market. These modems take advantage of the fact that, except for that last bit between your house and the telephone company's local office, most of the telephone network is digital. So the connection between your ISP and your telephone jack is digital almost all of the way, and those smart modem engineers have figured out a way to make the downstream data (that is, the data coming from the Internet to you) cruise through those digital lines and keep on cruising through that last analog bit at speeds approaching 56 Kbps.

Sounds great, huh? Well, it should be, but there's a catch — three catches in fact:

- First, this higher speed is *asymmetric*. In other words, the data only goes faster one way. The upstream flow maxes out at 33.6 Kbps — the same speed limit you have for regular modems. That's fine for Web surfing, where how fast data gets to you is what counts. But for Internet telephony and video, you'll be sending data back upstream, and the new modems won't do that any faster than the regular ones.
- Second, the high speed downstream flow doesn't work on all phone lines. With the higher speed comes even more sensitivity to the quality of your local phone lines, so you may not get a significantly higher downstream data rate than you would have with your old-fashioned 33.6 Kbps modem. We expect, however, that as the technology matures, this will become less of a problem.
- Lastly, no one standard method exists for squeezing the 56K from your modem connection. Right now, you can choose from two competing, incompatible systems: the Rockwell/Lucent 56K system and the US Robotics X2 system. If you buy the wrong one, you could end up with a modem that's not compatible with whichever system your ISP chooses to use. Again, however, we think that this will be less of a problem in the future, as the systems become compatible, or when one or the other becomes the standard.

Sounds like we're down on the new modem technology, doesn't it? Well, we're not. We just want to let you know that buying one of these modems now is a bit risky (for the reasons we mention in the preceding paragraphs). However, these new modems don't cost much more than the 33.6 Kbps ones, and they do make a lot of other 'Net activity — like Web browsing, file downloads, and checking e-mail — appreciably faster. So get one if you like; just don't expect too much of an increase in performance for Internet telephony and video applications.

Thou art, UART

Many older, and quite a few more recent, PCs have a bottleneck in the system that can keep you from getting the most from your modem. The chip that controls the serial port is called the *UART* (*U*niversal *A*synchronous *R*eceiver/*T*ransmitter). If you have an older UART chip, you may find that the stream of data flowing in and out of your modem can be too fast (if you're using a 28.8 Kbps modem) and the buffer that holds outgoing packets until they can be sent becomes overloaded. The result is lost data.

You can try two techniques to find out which UART you have. If you use Windows 95, go to the Start menu and choose Settings➪Control Panel. Double-click on Modems and select the Diagnostics tab. Then click on More info. The resulting screen shows you, among other things, which UART you have. If you have UART 16550AFN, then you are in luck. If you don't have Windows 95, you can check your UART from the DOS prompt, by typing MSD to open Microsoft Diagnostics. Select COM Ports, and the identity of your UART chip appears in the diagnostic window.

If you have an older UART, you can't do very much, short of buying a new Input/Output card for your machine, but at least you know.

Note: For one-way streaming multimedia applications, like real-time audio and video on the Web, these faster modems may indeed increase the audio and video quality — especially if the program you are listening to or watching is coming from a server with a fast Internet connection (and most are!).

If you want to get really smart about modems, pick up a copy of *Modems For Dummies,* 2nd Edition, by Tina Rathbone (published by IDG Books Worldwide, Inc.); it's a great source of information.

Fastest — ISDN or dedicated connections

Although by no means necessary to utilize Internet telephony and video programs, an ISDN dial-up connection or a high-speed dedicated Internet connection can certainly enhance your telephony experience. For example, these connection methods offer increased bandwidth. In addition, audio programs can use the highest-quality settings — the settings that use the least lossy compression techniques. And video benefits even more. ISDN speeds enable high-quality, full-motion videoconferencing connections and enable you to receive live or on-demand streams at the highest quality settings. (For more on streaming video, see Chapter 8.)

A dedicated connection typically gives you one more benefit: a fixed IP address. Chances are pretty good that, unless you're Bill Gates, you're not going to bring a dedicated digital line into your house, but if you have access to one at work or school, be thankful. You've got a good thing!

What about voice capable modems?

If you read any computer magazines or follow the computer news section in your newspaper, you may have heard of *voice capable* modems. Sounds like the perfect match for Internet telephony, doesn't it? Actually, no. Voice capable modems add a voice channel to your modem connection, so you can talk while transferring data.

This feature is useful if you are making a modem-to-modem connection with someone. For example, you can call a friend's or co-worker's computer directly (over the telephone network) with your modem, transfer files or other data, and also talk using a telephone plugged into the modem or with special software using your computer. This process is an example of *computer telephone integration*, or CTI.

In Internet telephony, however, your voice is the data! The connection is not over the telephone network, but over the Internet. So while voice capable modems are pretty neat, they don't have any real benefits over regular modems for Internet telephony. And they cost more.

Note: If you are going to use ISDN for your 'Net access, you need to skip past the modem section at the computer store and head for the NT1 and TA section. These two devices (often combined in the same box) are the ISDN equivalent of a modem. In fact they are often referred to as an ISDN modem, although technically they are not modems.

- **The NT1 is the Network Termination device** — it provides power and other network functions to the line. You really don't need to know what else it does, except to remember that you need power from the wall to run the NT — so if the lights go out, so does your ISDN line.
- **The TA is the Terminal Adapter** — it performs the modem-like function of transferring data from the computer onto the ISDN line (and vice versa).

Remember that a modem converts the digital signals used by computers to analog ones for transmission over phone lines. ISDN lines are digital, so no modulating or demodulating is going on.

You need to shop around to find the right device for your needs. Here are a few of the most popular options:

- Ascend Pipeline
- Farallon Netopia ISDN Modem
- Motorola BitSURFR
- US Robotics Sportster ISDN128K

Remember, not every TA works with your ISP's equipment, so check with your ISP to see what it supports before buying your TA. Some local phone companies are beginning to bundle TAs with ISDN service, and we expect that before too long, you can buy your TA, order an ISDN line, and get hooked up with an ISP all at one time.

Visit Dan Kegel's ISDN page at `http://www.alumni.caltech .edu/~dank/isdn/` as one of the best ways to find out what you need to know about ISDN.

Microphones and Speaker Options

You wouldn't be able talk into a standard telephone without a microphone and or listen without a speaker (in the handset or speakerphone). With your computer as a phone, the situation is the same. At a bare minimum, you need a microphone and a speaker to communicate. Many of today's multimedia-ready computers come with these items, but if yours didn't, you can get them easily and cheaply.

Here's talking to you, kid

The microphones are a good place to start, as they are rapidly becoming a core part of computers. Have a laptop? You probably have a built-in microphone. Buying a multimedia machine? You may find a tall, skinny-looking mike as part of that setup, too.

Many sound cards come bundled with a microphone in the box, as do some of the shrink-wrapped store versions of Internet telephony and video programs. These bundled microphones are usually cheap, but they do work. If you don't want to get too fancy, choose one of them. They're perfectly adequate.

To get the most comfort out of your Internet audio application, consider checking out some of the specialized microphone hardware that has been designed expressly for these situations:

- **Andrea Electronics ANC-100:** This model uses an active-noise cancellation circuit and is worn on the head. Other models can be used as either a desktop microphone or as headphones. For more information check Andrea's Web site at `http://www.andreaelectronics.com`.
- **InterActive SoundXchange handset:** Wanna try Internet phoning but still feel better with that old telephone handset wedged between your shoulder and ear? This product should make you happy. The Interactive SoundXchange from InterActive, Inc. looks and feels just like a regular telephone. (See Figure 4-1.) Plug it into the wall and into the back of your computer, and you're ready to feel comfortable.

Figure 4-1: A phone for your computer — the Interactive Sound-Xchange.

The sound quality is excellent, and you can adjust the volume without having to open any control panels on your computer. Several models are available, with both half- and full-duplex capability, and some even have built-in sound cards. (What's a sound card? Read Chapter 5 to find out. Mac users are excused from this one.) For more information on the SoundXchange, go to InterActive's Web site at `http://www.iact.com/`.

Macintosh computers use two different kinds of microphones, so make sure that you buy the right one for your machine. Most of the newer machines (like the Power Macintosh models) use a powered microphone called the Apple PlainTalk mike, which is different from the Apple mike. If you do not know the difference, the PlainTalk mike is sort of squished down and oval, while the Apple mike is circular.

What did you say?

Speakers are available in a wide range of prices and quality — from $20 battery-powered models to fancy three-piece units with subwoofers and high-powered amps. Which you choose is a matter of personal preference and budget. Just get the best speakers you can afford that fit on your desk.

The main decision you must make when choosing speakers, besides price, is whether to buy powered or nonpowered speakers. Powered speakers have an amplifier built into the speaker, while nonpowered speakers rely upon the computer's sound card for their amplification. Most of the time you either need, or get better results from, powered speakers.

If you use powered speakers, experiment with turning off your sound card's internal amplifier — check in your sound card's documentation for instructions.

Here are a few speaker models that we like:

- Sony SRS Series
- Yamaha YST Series
- Altec Lansing Multimedia
- AppleDesign

If you have one of those new A/V monitors, you probably already have built-in speakers. These will work fine to get you started.

Only one problem exists with the "hands-free/head-free" setup of a stand-alone microphone and external speakers: If you're running in a full-duplex mode (that is, sending and receiving audio at the same time), you can get feedback. The microphone not only picks up your voice as you speak, but it also picks up the sound coming out of the speaker and plays it back to the other person you're talking to. Feedback is annoying and disruptive. One solution is to use the InterActive SoundXchange handset approach just mentioned. You may also want to try some of these options:

- **Jabra EarPhone:** This neat piece of equipment is available from a company in San Diego. You can check out the Web site at `http://www.jabra.com`. The EarPhone is a one-piece, in-the-ear microphone and headphone combo that plugs directly into your computer. Both PC and Mac versions are available. The microphone is built into the external part of the headphone and offers much better quality than you'd expect.

  The EarPhone comes with neon rubber ear pieces of various sizes that fit in your ear and are actually quite comfortable (and they can be washed, which will probably make you feel better if all your friends want to check out Internet telephony on your machine). The PC version is even available with a license for VocalTec's Internet Phone included, so you can register your downloaded version for free (well, you paid for it in the price of the EarPhone, but the total price is cheaper than buying separately). The EarPhone also includes software that enables you to use your computer and modem as a kind of screen-based computer telephone for making calls over the PSTN.

- **Plain old headphones:** Another good option is to use some sort of headphone. Headphones eliminate feedback and give you clearer reception. A variety of headphones are available. Many Internet telephone users swear by the inexpensive in-the-ear models. You don't need stereo headphones — remember, voice over the 'Net, like voice over your regular telephone, is in mono.

In Chapter 20, we provide more tips for making yourself sound good.

Chapter 5

Getting Outfitted for Internet Video and Multimedia

In This Chapter

- Understanding hardware requirements for videoconferencing
- Ports versus cards — your choice
- What to buy?
- Video standards

If you thought Internet telephony was neat, get prepared for Internet video. Not only can you talk to people over the Internet, you can also see them, interact with them, collaborate with them, and more.

Your investment in Internet video can be extremely nominal — in some cases, less than $100 if you shop around. But the payoff can be great. Internet video works and is getting more sophisticated with each passing month. If you try it, you're almost certain to get hooked. You enter a whole new world through the Internet.

Getting Video on Your System

Adding video to your system is a fairly easy proposition. In fact, the basic concepts of Internet video do not differ much from those of Internet telephony.

Before we tell you about the Internet video hardware that's available on the market, you need to know a few things about video quality and how Internet video works. This chapter focuses on the hardware you need to get started with Internet video — Chapters 6 and 7 offer descriptions of the software for Windows systems and Macs, respectively.

Many of the concepts involved in audio also apply to video, except that a video signal is typically much larger than an audio signal. So to fit it within the same bandwidth as an audio signal, you must use even greater compression — or get a bigger pipe.

The moving image in a video stream is actually a series of still images, called *frames,* that are quickly sequenced one after another to simulate motion. Most of you are probably aware of how movies work, so we're not going to bore you here with any more basics. One thing to be aware of, however, is an index that is quite often referenced vis-à-vis video — frames per second (fps), which is an index of how often the image changes on your screen in one second. A good picture is about 30 fps; 10 to 15 is somewhat jerky; below 10 is considered almost a sort of slide show.

Many Internet video products hover in the 1 to 8 fps range for a variety of reasons. In fact, the size and quality of a video signal depends on several important factors:

- **The bandwidth available:** As we mention in other chapters in this book, the size of the pipe has a huge effect on quality. The bigger the pipe (bandwidth), the more frames you can send, which results in higher quality.
- **The nature of the image itself:** Better software applications, which use more sophisticated compression techniques, transmit only what changes in a picture. So slow-moving or still images are easier to compress because much of the image stays the same over time, resulting in higher quality. Fast-moving images, such as a video of a basketball game, take more processing power to compress, because a lot more information about the image changes, resulting in poorer quality.
- **Image size:** This falls into the no-kidding category. The larger the image, the more data is required to reproduce it. Image size is directly related to the next item — resolution.
- **Resolution:** In the computer world, resolution is determined by how many pixels are used to display an image. (The term *pixel* — short for picture element — is the smallest element that a device can show.) The more pixels used to create a picture, the more precise the resolution of the picture. The greater the resolution, the more data to be transmitted.
- **Refresh rate:** The refresh rate is the other major factor in image size. Refresh rates are calculated in frames per second. If an image is being refreshed at 30 fps, a great deal more data must be transmitted in the same amount of time than with a 15 fps image.
- **Color versus black and white:** A color image requires more data for each pixel than a black-and-white image because the computer must define the color and combine several different colors on the display to show it. The more colors you have, the more data that must be transmitted.

To get started with video, you need a camera, possibly a video board (depending on the type of camera you pick), and the software. You find out about video hardware and standards in this chapter; check out Chapters 6 and 7 to read about your software choices.

Here's Looking At You, Kid

The first and most obvious requirement is a camera. What kind of camera you add drives your requirements for any further hardware additions.

You can choose from three basic types of cameras:

- **Digital video output computer cameras:** Your PC is digital, so using a digital camera makes some sense. Digital cameras use something called a *Charge-Coupled Device,* or CCD (a chip in your camera that converts light into an electrical signal), to convert the analog images entering the camera into a digital signal that your computer can understand.

  With a digital output camera, you can plug the camera into the back of your machine, and you're ready to go. Although more models are likely to enter the market in the future, for now the Connectix QuickCam and Color QuickCam stand alone. And with its mere $100 price tag (for the black-and-white model) you would almost be crazy to want anything else. More on the QuickCam in the following section.

- **Analog video output computer cameras:** Analog cameras do not output their video signals digitally, so they need to be helped along by a digital-converting video card that slides into your computer. These cameras, like the digital video output computer cameras just mentioned, also use a CCD to convert the light (by which we mean the image) into a digital signal, but then they go an extra step and convert the digital signal put out by the CCD into an analog video signal.

  The video capture card converts the signal back to digital so that your computer can accept the images. (Why it goes from analog to digital to analog to digital is one of those mysteries we are better off not understanding.) We discuss video capture cards further a little later in this chapter.

- **General-purpose camcorders:** Yes, these are standard, run-of-the-mill, *America's Funniest Home Videos*-types of video camcorders. General-purpose video camcorders output analog video signals, so you again need a video capture card to digitize the signal for your computer.

  With that video capture card, however, you can plug your camera into the back of your computer to broadcast camcorder images over the 'Net. Want to show your mom your kids' latest mess? Just plug your camera in and show it to her with Internet video.

We've just skimmed over your camera options for Internet video. Read on to find out more specifics.

Digital video output computer cameras — the Connectix QuickCam and Color QuickCam

The QuickCam looks strangely like a cross between a large golf ball and some sort of floating eyeball from a science fiction movie. It is a digital camera, which means that it captures the image and outputs a digital signal that is ready for your computer to process.

Because of immediate digital output, you don't need any additional hardware to use the QuickCam. All you need to do is plug it into the parallel port of your PC or into the printer or modem (serial) port on the back of your Macintosh, and you are ready to go.

The QuickCam is

- Inexpensive (about $100)
- Easy to set up and use
- Equipped with a built-in microphone (Mac versions only)
- A device of fairly good quality

The built-in QuickCam microphone for Macs provides a lower level of quality than the Apple microphones do, and using the QuickCam microphone reduces your frame rate. We recommend that you use an external microphone plugged into your Mac's microphone jack.

Because the QuickCam is so quick, easy to set up, and inexpensive, it's probably your best bet for getting started in Internet videoconferencing. It does have the following disadvantages, however:

- It doesn't support color; 64 shades of gray is it.
- It doesn't have as high a resolution as some more-expensive cameras.

The Color QuickCam will set you back a little more than twice as much as the black-and-white version and has all of the same advantages, plus:

- You get your picture in living color (hope that's not a trademark!)
- A manual focus that expands the range of focus (you can get a lot closer to the camera)

Parallel universes

The QuickCam family of products has been an overwhelming success in the computer marketplace. You really have no easier or cheaper way to get video into your PC or Mac. If you remember Econ 101, you won't be surprised to find out that other companies are looking at ways to get into this field. As we hear it, a lot of consumer electronics companies (who already know how to make camcorders) are burning the midnight oil developing their own answer to the QuickCam. One example of this is Vivitar, the camera and copier and bunches of other stuff company, which has released a parallel port camera called the MPP-2 (Motion Picture Phone). You can get more information on the MPP-2 at the Vivitar Web site, `http://www.vivitarcorp.com/MMP2.html`.

Right now, the QuickCam family has no real competition in the low-cost digital camera arena. You can expect to see some of the major consumer electronics firms (which make a great deal of money selling camcorders) playing on their experience and offering their own alternatives soon.

Parallel port cameras like the QuickCam use your computer's CPU to digitize your video, while separate camera/video card combinations offload this task to chips on the video card. If your PC doesn't have a whole lot of extra CPU horsepower available (for example, if you have a 486 or a slower Pentium), you may get better results with the more expensive camera and video card combination.

Most mail-order catalogs sell the QuickCam. For more information, go to the Connectix QuickCam Web site at `http://www.quickcam.com`.

Analog video output computer cameras

If you're willing to spend a little more money for higher quality, you can buy a dedicated color camera designed purposely for videoconferencing. These cameras may cost anywhere from $300 to $1,000 or more and require a video capture card. So these cameras require a bit more of an investment than the QuickCam, but they offer higher image quality and color.

The following sections describe a few of the more popular models.

VideoLabs FlexCam

This desktop camera is mounted on a long stalk and a sturdy base that sits on your desk. It includes the following features:

- Color CCD
- Built-in stereo microphone
- Swivel head for aiming the lens
- Flexible stalk for positioning

The FlexCam runs about $595, and you can get it from major computer catalogs. For more information and other places to buy a FlexCam, go to the VideoLabs Web site at `http://www.flexcam.com/`.

Sony PC Cam

The PC Cam is actually quite similar in appearance to the FlexCam and includes the following features:

- Built-in microphone
- Color CCD
- Flexible stalk and swivel head similar to FlexCam

The PC Cam runs about $499. For more information and a list of dealers, go to the Sony Electronics Web page at `http://www.sel.sony.com/SEL/ccpg/cav/computer/pc1.html`.

Winnov VideumCam

The Winnov VideumCam is a new product with a price much closer to the QuickCam than most of its competitors — about 200 bucks. Of course, you still need to buy a video capture card — but the total price may be only twice as much as a Color QuickCam, instead of three or four times as much. The VideumCam's features include

- Color CCD
- Built-in microphone
- Draws power from its companion video card (the Videum AV, which we mention in the following tip), so you don't need to plug it into a wall outlet or power strip

The combination of a VideumCam and its companion video capture/audio card, the Videum AV, is an increasingly popular alternative to the QuickCam. For about $360, you get the whole package — still more expensive than the QuickCam, but you do get a higher quality picture and less of a load on your PC's CPU.

Is NTSC your PAL?

When you enter the world of video, you often hear terms such as NTSC, PAL, and SECAM being thrown about. These terms refer to the analog video signal standards used for television around the world. In the U.S., NTSC (for National Television Standards Committee) is the bottom line for television. This is the video signal that is transmitted over the air, through your cable TV, and out the back of your VCR. NTSC, PAL, and SECAM actually have very little to do with digital video, which uses its own standards (see the section discussing video standards, later in this chapter), but they are the signals that flow out of your video camera and into your video card.

By the way, NTSC is used throughout North America and in many parts of Asia and South America. PAL and SECAM are standards in other parts of the world.

The S-Video (also called S-VHS) standard is a variation of NTSC (with a slightly higher resolution) that is used by some video cassette recorders. The main difference to you is that it uses a different jack plug, so if you have S-Video coming out of your camcorder, make sure that you select a video capture board that can handle it.

A great source of information and prices of digital and analog output computer cameras is Picturephone Direct. Visit its Web site at `http://picturephone.com`.

Camcorders and general-purpose video cameras

The third option for adding video to your system is to use a camcorder or dedicated video camera. These cameras add a video capture card to your shopping list but have the following advantages, especially relative to the QuickCam:

- You may already have such a camera at home.
- The work of digitizing your video is done by the video capture card instead of your CPU — so you have more power left over for other things (like running the videoconferencing program).
- Most models have a higher resolution and, therefore, a sharper picture than a QuickCam.

The following disadvantages exist if you go this route, however:

- In addition to the expense of the video capture card, the cameras generally cost more.
- The cameras add complexity in setting up your system.

You can use just about any camcorder as long as it has an NTSC-composite or S-Video output that you can plug into your video capture card.

You need to decide for yourself whether a more sophisticated system involving a general-purpose camera and a video capture card is worth the expense to you. If you already own a camcorder, the combination may be a good way to use what you have, but the expense is still higher than buying a QuickCam.

Given the current state of the art in 'Net videoconferencing, we suggest that you use the QuickCam to get yourself up and going. Unless you also use the camera for other multimedia purposes, such as creating movies or doing high-speed LAN or ISDN-based videoconferencing, the difference in quality is too low to make the extra cost and complexity of a camera/video card combo worthwhile.

Video capture cards

A *video capture card* is really just the video equivalent of a sound card. Its sole purpose in life is to take incoming analog video signals and convert them into a digital format that can be read by your computer. Prices for capture cards are higher than those for sound cards and usually range between $250 and $400, depending on the model.

Like sound cards, too many brands and models of video cards exist to list here. Some of the most popular cards are

- Creative Labs FS200 and RT300
- Digital Vision ComputerEyes 1024
- Logitech Movie Man
- Winnov Videum AV
- VideoLabs T320

Good sources of information on video capture cards include the Web sites of the videoconferencing software manufacturers that we list in Chapters 6 and 7. You also can check out a good list of available cards at the following address:

```
http://www.hertz1.com/videocap.htm
```

Note: You may already know that Macs don't need sound cards. If so, you're probably hoping that they don't need video cards either. Sorry. Unless you have one of the AV models that have built-in video support, you need to get a card.

Before you buy a video capture card, check the Web sites for the videoconferencing software packages you are considering. Find the FAQ (frequently asked questions) file, and you'll usually find a listing of hardware that has been tested and is *known* to work with the product. Better safe than sorry.

Taking a Look at Video Standards

You've probably already guessed that standards are pretty important for how things work on the 'Net. That's why we mention them throughout this book. We're not trying to bore you — it's just important that you have at least some idea about them, because as you dig further into these subjects, the standards come up quite often.

True Motion video standards

The True Motion standards govern how video is stored and played back digitally — meaning on your computer, although these standards are becoming increasingly important to your television set as it gets more sophisticated. (In fact, one of the issues being raised by future-looking pundits is whether the TV will become more sophisticated and replace the PC, or vice versa. We think that they both will become more capable, but both will continue to exist independently for quite some time.)

Probably the most important True Motion standard is MPEG (for Motion Picture Experts Group, the folks who created it). You may be familiar with this term, either from downloading MPEG files to play back on your computer or because a variant of it is used by the popular DirectTV mini-satellite system. If you dig even deeper into MPEG, you discover variations of the standard — MPEG1, MPEG2, and so on. These are successive generations of the standard, renamed as it is refined and updated by the MPEG group. At this time, MPEG2 is the latest standard, but MPEG3 and even MPEG4 are on their way. Don't pull out too much hair worrying about these; they should be invisible to you if you are using an MPEG application.

Videoconferencing standards

While we're talking about standards, we want to take a second to bump into our old friend, H.323. As we mention in Chapter 2 (you didn't skip all of that boring background information, did you?), the H.323 standard defines how computers conduct multimedia conferencing (audio and video) over networks such as the Internet. A big part of that standard is the video codecs, which make your video signal suitable for transmission over the Internet — H.263 is the one you see used most often.

In many cases, the product literature for video hardware mentions compatibility with other standards. These standards don't have all that much to do with Internet telephony and videoconferencing, but you can impress your friends if you know what they are:

- H.320 is the standard for videoconferencing over ISDN and similar circuit-switched digital data lines. Much of the development of H.323 was done with H.320 in mind, in an effort to allow compatibility between different conferencing systems.
- H.324 is a standard for videoconferencing over regular phone lines (POTS, or Plain Old Telephone Service). Many of the popular 'Net videoconferencing programs, such as Intel Internet Video Phone, have a companion product that utilizes much of the same technology to allow point-to-point videoconferencing over a standard phone connection.

In the not-so-distant future, you'll probably see companies develop gateway products that allow different H.32*something* products to connect together. For example, an H.320-to-H.323 gateway may let a remote user plug into an ISDN videoconference from home via an Internet connection.

QuickTime and Video for Windows

Two other industry standards come into play in the realm of computer video. QuickTime is the Apple video technology and is used both on Macintosh and Windows computers. QuickTime is basically a technology for recording and playing back video on your machine. Video for Windows is Microsoft's answer to QuickTime. If these programs aren't familiar to you, don't worry. Just remember that these are underlying technologies that enable your machine to use video.

Chapter 6

Getting Your PC Set Up for Internet Telephony

In This Chapter

- Meeting the minimum requirements to use Internet telephony on a PC
- Choosing the best PC for Internet telephony
- Choosing the best PC programs
- Installing PC telephony programs

You probably have an idea of the peripheral devices that you need for Internet telephony and video — things such as modems, speakers, and microphones.

However, to make these programs work on a specific platform, you need some additional hardware and software preparation to be 100 percent ready to hit the 'Net.

This chapter concentrates on PC systems; the following chapter focuses on Macs.

PC Minimum System Requirements

Before we get into the nitty-gritty details of what your system needs to take part in this Internet revolution, here is our idea of what, at a minimum, your system should include:

- 486 PC
- Half-duplex 16-bit sound card
- Windows 3.11 or Windows 95
- Winsock 1.1 or compliant
- 8MB RAM

PC Recommended System Requirements

The preceding section lists the bare-bones minimum system requirements for getting started with Internet telephony and video. However (and this is a big however), we recommend that, to really get the most out of what Internet telephony and video have to offer, you outfit yourself with the following:

- Pentium or fast 486 (at least 66 MHz)
- Full-duplex 16-bit sound card
- Windows 95 (includes Winsock 1.1)
- 16MB RAM

More on What You Need to Make Your System Work

In the following sections of this chapter, we go through each of the system requirements in a little more detail. If your system is deficient in one area or another, check out that particular portion of the chapter and then feel free to skip to the second half of the chapter, which gets into IBM-compatible telephony programs available today.

486 or Pentium processor

All software has a minimum configuration under which it can run, but the *recommended configuration* supports ideal operation. Internet telephony and video programs are no different.

Some telephony programs tell you that you can use a 386 machine, and that advice is probably true. Read further on the same page of the manual or Web site, however, and you're bound to find that the developer *recommends* that you use at least a fast 486 (DX33 or faster).

Chapter 2 notes that the software handles signal compression and encoding. Consequently, your CPU is responsible for running the algorithms that transform your analog voice into something that can be sent out over the 'Net digitally. The higher-quality algorithms put a bigger load on your CPU, and a slower machine does not have enough horsepower to handle the load.

We have tried using Internet telephony software on a range of systems. Yes, many programs do work on 386 machines, but they run far better on a 486 or a Pentium. In general, 386 machines have a harder time keeping up with the requirements of new computer software, so this result does not come as a surprise.

We typically use a 486 DX4 100 MHz machine in our office for testing, and we find that it performs well. But you're even better off with a Pentium, particularly if you're buying a new computer. If you're still using a 386, go ahead and try some of the free demo versions of telephony software, but hold off on buying software until you upgrade your machine to a newer processor. You'll be happier with the overall results.

8MB RAM

The 8MB figure is an important number, as almost all the programs require or "recommend" at least this much memory. Even better, though, is 16MB, especially if you take into account the fact that you may want to run other programs, such as Netscape, simultaneously. If you don't have enough physical memory installed, Windows uses the hard drive to create virtual memory, but this process can and does slow things down. If you're running memory-intensive programs — such as Microsoft Word or Excel — you're likely to want to add memory anyway for reasons aside from Internet telephony.

Windows 3.1 or later

The programs we discuss in this chapter are not DOS programs (thank goodness!). Most telephony programs require at least Windows 3.1 or a later version of Windows, and many now have 32-bit versions available for Windows 95. Using Windows 95 provides many benefits, not the least of which is (relative) ease of setup. As you see, we used Windows 95 to set up and run the programs that we discuss as we go along in the book.

If you haven't yet updated to Windows 95, you really should. The Windows 95 programs for Internet telephony are more common, have more features, and generally work better.

TCP/IP — Winsock 1.1 or better compliant

As we discuss in Chapter 2, the Internet is a TCP/IP-based network. For your computer to talk over this network, you need two things. The first, as you may already know, is some sort of an Internet connection, either dedicated or dial-up — Serial Line Interface Protocol (SLIP) or Point-to-Point Protocol (PPP), remember? The second item is *Winsock*, which stands for *Win*dows *Sock*et. Winsock is a protocol that provides the interface between the program your computer is running and the Internet.

We have a reason to print two warning icons for this section. If you have a problem in your configuration, it's going to be with the Winsock program. You see, several versions of Winsock are available, and this variation poses the challenge of making sure that you have the right version to match your setup. The problem is that Winsock is often a tricky item to troubleshoot, because wrong or incompatible versions can cause intermittent problems in actual program use.

Whole books have been written about this subject, and you can read them if you'd like. What you need to know for sure is that you have, or can get, a Winsock version 1.1 or a Winsock that is compliant with version 1.1. Windows 95 comes with its own version of Winsock already installed for you. If you're using an older version of Windows, you need to get a copy of Winsock and install it. See the sidebar "Plugging into (Windows) sockets."

Some of the programs we discuss here require you to be using the Microsoft Winsock that comes with Windows 95. We like that one best of all anyway and highly recommend that you use it.

Sound cards

Unless you have a newer multimedia PC, you probably need to purchase a sound card of some sort.

Until recently, most PCs came without any sound support — other than an internal beeper that notified you as you're restarting or doing something you shouldn't. (You've probably heard that metallic *boing* more times than you can count.) To hear voice, music, or just more interesting noises, you need to plug in a sound card.

Luckily, if you bought your computer recently (in the last year or so), it probably came out of the box with a built-in sound card.

Plugging into (Windows) sockets

Winsock is a critical piece of software that enables your computer to establish and maintain a connection over a TCP/IP network (the Internet, to be specific). PC programs that run over the Internet, such as Internet telephony programs or Web browsers, are known collectively as Winsock applications.

If you're using Windows 95, you already have a Winsock built into the system software. Users of Windows 3.11 must fend for themselves. Here are a few suggestions for obtaining Winsock:

- Purchase a combination Internet set-up package, such as the Spry Internet in a Box or the Netscape Navigator Personal Edition, which includes Winsock.
- Get Winsock from your ISP. Most ISPs include a free or shareware Winsock version, such as Trumpet Winsock, in their start-up software packages.
- Use a telephony application, such as WebTalk, that includes Winsock in the software.

  Note: We prefer to use Windows 95 because it facilitates dial-up 'Net access. Its Winsock version is already installed and ready to go!

Winsock is supposed to be a standard. That is, all Winsock programs should be able to work with your Winsock 1.1-compliant Winsock (what a mouthful). Some programmers have decided that they can do it better, so they create their own specialized Winsock. We recommend that you don't install any of these hybrids, but if you do, make sure that you back up your original Winsock so that you can reinstall it if everything comes crashing down.

8-, 16-, and 32-bit implementations

Depending on how much you want to spend (and it can vary greatly — from $100 on the low end up to $400 or more for a fancy high-end card), you can pick cards that implement sound in 8, 16, or 32 bits.

The word *bit* in the term *8-bit sound card,* for example, refers to the amount of the data sampled by the sound card. (We discuss sampling in Chapter 2.) In general, the bigger the size of the sample, the higher the sound quality.

What does this tech talk mean to you? Here's the scoop:

- 8-bit sound cards (which are getting harder and harder to find) are basically game cards, designed to play back lower-quality sound (sort of like an AM radio).
- 16-bit cards are the norm and can be purchased relatively cheaply. The Soundblaster 16 that we use on our test machine costs only 100 bucks; it provides a sampling rate equivalent to a CD player (44 kHz) and can give you very good sound quality.

- 32-bit cards are the fancy sound models that provide high-end sound manipulation. They support great sound and can do other neat stuff that you never use in the realm of Internet telephony. Unfortunately, these cool features (such as MIDI and synthesizer stuff) add to the price tag.

Get yourself a 16-bit sound card. Don't skimp on a cheaper 8-bit model; it's just not worth it. On the other hand, 32-bit cards are overkill for Internet telephony.

Full versus half duplex

Most people, when they hear the word *duplex,* probably think of something you live in. We're not talking houses here.

Full duplex versus *half duplex* refers to the expected path of the sound going through your sound card.

Half-duplex sound works like a walkie-talkie or a CB radio. Only one person at a time can talk. You know: "Roger, over and out." (By the way, if you're ever really using a radio around people who know what they are doing, never — never — say "over and out." It's either *over* or it's *out*, not both at the same time.)

Full duplex is like a real-world conversation or one on a regular telephone. Both parties can talk and hear each other at the same time. Unless, of course, one of you is a lot louder than the other. . . .

Full duplex is almost required for your Internet telephony needs. All these programs *work* with half-duplex audio, but trust us on this, you will be much happier in full duplex. Conversations flow much more smoothly and feel more natural. Find a card that supports full-duplex audio. (So many are available that we don't even try to list them.) The extra cost of a full-duplex sound card is probably less than tens of dollars — and well worth the expense.

We don't want to contradict ourselves, but . . . some instances do exist where you may end up preferring a half-duplex conversation. This situation occurs most often if you're talking to someone halfway around the world and have an excessive amount of lag or delay in your conversation. In this case, you may want to switch to a half-duplex mode to avoid confusion.

A sound card sounds like just what I need

So you need to buy a sound card, but you don't know where to start looking. Fortunately, a few good sites on the WWW can help you. First look on the sites of the Internet telephony program makers themselves (we list the most popular ones later in this chapter). Another good source is the Hertz Technologies group at `http://www.hertz1.com/fulldup.htm`.

VGA card of 256 colors

Many telephony programs require 256-color VGA cards and monitors to appear correctly on your screen. If your computer can run these programs, however, you probably already have at least this level of graphics capability.

Older PCs

If you have an older PC and you want to run PC-based Internet telephony and video programs, sorry, you're out of luck! We don't want to get you down, but if you have an older machine, such as a PC XT or some sort of 286 machine, you really do need to do some upgrading. We doubt that we're the first to tell you this. You need to buy a new machine, preferably a Pentium, and start over again. The amount of money you would have to spend to upgrade your CPU, RAM, hard drive, and so on just wouldn't be worth it.

If you have a fairly modern 386 machine with sufficient RAM and room to plug in the sound card, you can give an upgrade a try; but we still don't recommend it. Upgrading the CPU in some 386 models is possible. However, you should look long and hard at the cost and benefits of upgrading versus buying a new machine. By the time you mess with it all, you probably are not that far away from the cost of a brand-new machine (which comes with a warranty and looks all shiny on your desk).

An Introduction to PC Internet Telephony Programs — WOW!

One of the strengths of the PC is the wide variety of software available for nearly all applications, including telephony. (Just compare this section to the one where we review Mac stuff.) In fact, your biggest problem isn't finding software you like; it's narrowing down the list to a reasonable number to try.

In the next few pages, we help you through the maze by describing the most popular programs. Remember: Things are changing fast. New programs come out, names change, and big companies buy little companies. Don't be surprised if something you've heard of is not mentioned here — or if you pick a program from this section and can't find it.

We just don't have room here to list all of the programs you can find to do Internet telephony and videoconferencing on your PC — there are just too darn many. So, instead we present a highly subjective and arbitrary listing of those programs that are most popular, widely used, or just worth your while to check out. Don't let the exclusion of any particular program here keep you from trying it out — we simply had to draw a line somewhere or be forced to present you with a multi-volume *Internet Telephony For Dummies* that would rival an encyclopedia (or the manual from your word processor) for shelf space.

Microsoft NetMeeting

A relative newcomer to the Internet telephony marketplace, Microsoft has nonetheless taken the 'Net world by storm with NetMeeting. A free program (hmmm, we just want to say that again, it sounds so good — free!) which is part of the Internet Explorer product line, NetMeeting is a fully compatible (both H.323 and T.120 — refer to Chapter 2 if these terms mean nothing to you) Internet conferencing program for Windows 95 and NT 4.0. With NetMeeting, you can conduct audio and video conferences with a single party or conduct data conferences with several people at one time. While in a data conference, you can talk to everybody within the conference via text chat, or if you wish, you can pair off with one other person in the conference and share video and audio.

NetMeeting's telephony features include

- Color videoconferencing
- Full-duplex audio with multiple codecs for optimizing audio quality
- Directory server for placing calls by entering an e-mail address (the Internet Location Service, or ILS)
- Direct calling to an IP address
- A speed dial "phone book" that utilizes the ILS to tell you if your friends are currently online
- The capability to call regular telephones through an H.323 gateway

The data conferencing features include

- Text chat that lets you type messages to others
- A whiteboard for sharing and viewing drawings, pictures, and screen captures
- File transfer to send or receive files from others in the conference
- Application sharing, which lets participants in a conference actually edit and manipulate files in just about any Windows program — even if they don't have that program on their own PC

As we mention earlier, NetMeeting is part of the Internet Explorer family. This means that it has a high degree of integration with other Internet Explorer programs. For example, you can find listings of NetMeeting users on the Web and connect with just one mouse click in Internet Explorer. Or if you can't connect via NetMeeting, you can send a message to someone by using Internet Mail and News (the e-mail and newsgroup component of Internet Explorer).

More information and a free download of NetMeeting are waiting for you at `http://www.microsoft.com/netmeeting/`. Chapter 9 goes into more detail on using NetMeeting.

Netscape Conference

Netscape used to have the Internet software market all to itself — when you said browser, you really meant Netscape Navigator. That all changed when Microsoft got serious about the Internet and started putting its considerable weight behind products such as Internet Explorer. Now the two companies are in a pitched battle for the hearts and minds (not to mention wallets) of Internet users. So it should come as no surprise that just as Internet Explorer has a telephony component (NetMeeting), Netscape Communicator (the latest version of the Netscape Swiss-Army-Knife-style Internet tool) has one, too. Previous versions were known as CoolTalk, but the latest and greatest (beta version, as we write) is called Conference.

Like NetMeeting, Conference is an H.323-compliant 'Net telephony program with many multimedia collaboration features included. Also like NetMeeting, Conference is a free add-on to the browser package. You already know how we feel about free software. (We like it.)

Conference includes the following features:

- Full-duplex audio with selectable codecs
- Directory service for locating other users (this one is known as ULS, or user location service)
- Direct dialing capability (by entering an IP address)
- A phone book for storing frequently called numbers
- Text chat
- A whiteboard for sharing images and collaboration
- File transfer capabilities
- Voice mail
- A collaborative browsing feature that lets you lead other users on a tour of the Web by using Netscape Navigator

Like NetMeeting, Conference is well integrated with the rest of the Communicator family — it's really just another component on the Communicator task bar, just like the mail program or the newsgroup reader.

As we write this (late spring 1997), Conference has not been finalized, and a few elements of the H.323 standard have not yet been incorporated. In particular, Conference has no videoconferencing element, and it is not compatible with the T.120 data conferencing standard.

You can download Conference from the Netscape Web site at `http://www.netscape.com`. You can find detailed instructions for using Netscape Conference in Chapter 12 of this book.

Conference is not available as a separate download, as CoolTalk once was. You must download the entire Communicator package — and be sure to select the "full install" option when you do, or you won't get the Conference module with your Communicator download.

Intel Internet Video Phone

Besides making a huge chunk of the CPUs found in the world's PCs, Intel does do a few other things (as if the CPU business weren't enough). One of these "sidelines" for the world's biggest CPU manufacturer is the ProShare family of videoconferencing products — which are a market leader for desktop videoconferencing over ISDN and other high-speed data lines. The folks at Intel know a good thing when they see one — and they see the Internet as a good place to expand their product line. So they developed the Intel Internet Video Phone, which brings some of that ProShare technology to us 'Net users.

Internet Video Phone is an H.323-compliant product, so you can use it to talk to other telephony products (like NetMeeting and Conference) as well as other Internet Video Phone users. In fact, Intel worked closely with Microsoft to make the H.323 standard a reality, and was the first company to release an H.323-compliant product.

As we write this, Internet Video Phone is not yet a final product, but the beta version is available, and it works just fine. One neat feature of Internet Video Phone is a built-in version checker which automatically logs into Intel's 'Net site and verifies that you still have the most current version of the program. No longer will you have to constantly check into the Web site to see if a new version has been released — the program does this for you. Just the thing for all of you (or should we say, us) "gotta have the latest and greatest" 'Net junkies.

Other features of Internet Video Phone include

- Full color videoconferencing
- Full-duplex audio
- A "connection advisor" that keeps track of your computer and 'Net connection so you can make sure all is running smoothly
- Directory services (it uses the same system as NetMeeting) for locating other users
- Speed dial for frequently called parties
- Proxy support, which enables you (or your system administrator) to configure Internet Video Phone to connect through a firewall protecting your company's intranet

Internet Video Phone isn't designed to be a full-featured data conferencing program like NetMeeting, so you won't find features like whiteboards or file transfer functions. However, several third-party add-on products work together with Internet Video Phone to let you do more with the program — for example, share pictures or play games. You can find a listing of these programs, and the Internet Video Phone itself at Intel's "Connected PC" Web site at `http://connectedpc.com/`.

You can find out more about using Intel Internet Video Phone in Chapter 11.

Internet Video Phone requires a bit more CPU horsepower than the baseline system we describe earlier in this chapter. In fact, a minimum of a 90 MHz Pentium processor and 16MB of RAM is required to send and receive video. A less powerful Pentium can be used in an audio-only mode.

VocalTec Internet Phone

The grandfather of Internet telephony programs, Internet Phone, was developed by an Israel-based firm called VocalTec. (Its U.S. headquarters is in New Jersey, and it was recently listed on the NASDAQ stock exchange.)

VocalTec claims that its Internet Phone program has been downloaded from its Web site (`http://www.vocaltec.com`) several hundred thousand times. In fact, the program has reached the top ten and top five sales lists for PC communications programs, as listed in a number of computer publications. That's pretty impressive for a technology that barely existed a year ago.

Internet Phone connects calls by using Internet phone "servers," which are the basis of the Internet phone network. These servers are based on the Internet relay chat (IRC) server system that you may have encountered elsewhere on the Internet. After logging onto the server, you can find a list of chat rooms and, within them, a listing of people available to talk with. You can specify private topics that give only you and your friends (family, coworkers, and so on) access, so you can quickly find those with whom you would like to talk. Internet Phone also includes phone book features and the capability to place a call directly bypassing the server system by using a directory service or dialing an IP address directly.

The Windows 95 version of Internet Phone (version 4.5) also includes a color videoconferencing capability — you can send and see video while talking with other users of the Windows 95 version. Sorry to tell you that the 16-bit Windows version can neither send nor receive video.

Other features of Internet Phone include

- Full-duplex sound support (depending, of course, on your hardware configuration)
- Configurable audio quality modes
- Whiteboard, which you can use to share images and drawings with other people online
- Text chat tool that allows you to type messages back and forth
- File transfer tool that allows you to send data files while you are talking
- Voice mail — leave messages in your own voice for people with whom you can't connect
- Capability to be connected to more than one party at once — although you can conduct a voice conversation with only one person at a time

You can download the latest demo version of Internet Phone from the VocalTec Web site for free. This version can be used for seven days. VocalTec is constantly updating Internet Phone, so don't be surprised if more new features are available by the time you download the demo. At the end of this chapter, we show you how to install Internet Phone.

If you're interested in cross-platform compatibility, VocalTec has a Mac version of the Internet Phone, as well. The Mac version lacks many of the features of the Windows version (stuff like video, whiteboarding, voice mail, and text chat), but it is fully compatible with the Windows version when it comes down to the bottom line — audio.

Several hardware manufacturers, such as Jabra and Interactive (we mention them in Chapter 4), are bundling Internet Phone with their products, so if you're buying some new hardware, you may get a fully functional version of Internet Phone for free. What a deal!

Internet Phone is not yet compatible with the H.323 standard, so if you purchase it, you'll be limited to talking with other Internet Phone users only. Luckily, a whole bunch of them are out there.

We discuss using Internet Phone in more detail in Chapter 10.

CU-SeeMe

CU-SeeMe is the original Internet videoconferencing program. First developed as freeware by a team at Cornell University (the CU part of the name), a commercial version has been released by White Pine Software in New Hampshire. The commercial version costs a nominal amount, about $69.

Both versions are available for you to use: You can find the latest freeware version at `http://cu-seeme.cornell.edu`, while a 30-day demo version of White Pine's CU-SeeMe is available at `http://www.cu-seeme.com`.

CU-SeeMe is available for Macs and PCs — and the two versions are compatible so you can conference with both Mac and PC users. CU-SeeMe enables you to connect to another person either directly, by entering an IP address, or to a multiparty conference by connecting to a reflector (server) site.

A *reflector site* is a location on the 'Net that runs a special UNIX or Windows NT CU-SeeMe program (Reflector). This program reflects signals back to those who are signaling with that site so that multiple parties can log onto the reflector and participate in conferences. You find out more about reflector sites in Chapter 13.

Because it was designed with 10-megabits-per-second LANs and campus backbones in mind, the original CU-SeeMe doesn't work very well with the low-bandwidth modems that many of us have plugged into our computers at home. The bandwidth is just not adequate. This problem isn't really too bad with the video transmission, but the audio codecs in the freeware version of CU-SeeMe aren't really designed to work at all on a modem connection.

Fortunately, that problem is being addressed by the developers of the commercial version. Enhanced CU-SeeMe offers color capabilities and is designed to run well over lower-bandwidth Internet connections, such as a 28.8 Kbps modem connection, as well as over higher bandwidths as with the original program.

If you are using the Cornell University freeware version of CU-SeeMe, you can't see the video of people who are sending color video with the Enhanced version of CU-SeeMe.

Both versions of CU-SeeMe enable you to do the following:

- Participate in a video conference one on one.
- Team with as many as 8 active participants (12 in the White Pine version).

  Note: More than 8 participants can be on a CU-SeeMe reflector site at once, but you are limited to opening only 8 (or 12) video windows at a time. You can easily close some of your open windows to see others, but you can never have more than eight open at a time.
- Participate in an audio-only mode.

You can always use an Internet videoconferencing program for audio-only conferencing if you want or need to. This is particularly effective if you have a slow (dial-up) Internet connection.

With CU-SeeMe, you can also do the following:

- Watch broadcasts (one-way video).
- Share whiteboards or text chats.

  Note: A *whiteboard* on your computer is just like a whiteboard in your office or classroom. You can draw on it, put text on it, or even paste things (such as pictures) on it.
- Just *lurk* at a public reflector site (watch the video, but not send).

Features of Enhanced CU-SeeMe include the following:

- Caller ID
- Password security
- Selectable audio codecs (so that you can adjust audio compression based on your bandwidth availability)
- Color video
- Phone book for people and reflector sites

White Pine's version of CU-SeeMe used to be called Enhanced CU-SeeMe, but with the new version (still in a beta-testing phase as we write), the name has been changed back to just CU-SeeMe. Just to confuse matters a bit more, the Mac version of White Pine's software is still called Enhanced. Go figure.

CU-SeeMe is not yet H.323 compatible, so you won't be able to communicate with anyone using a different program. However, White Pine has just announced a new conference server program, called MeetingPoint. When this product becomes available, it will allow NetMeeting and Intel Internet Video Phone users to log into the conference server (an updated version of a reflector) and share audio and video with CU-SeeMe users.

You find out more about using CU-SeeMe in Chapter 13.

WebPhone

Itelco (Internet Telephone Co.), a division of NetSpeak Corporation, offers WebPhone. WebPhone has a neat, cellular-phonelike interface with four lines. This feature enables you to put one line on hold to answer other lines — you can even connect more than one line together for a conference call.

WebPhone uses a point-to-point calling scheme, with entry of an e-mail or IP address necessary to place a call. An online directory assistance server is available if you don't know someone's calling information.

Among the features of WebPhone are the following:

- Color videoconferencing using the H.263 standard
- Full-duplex operation
- Encryption of your conversation
- Call holding and blocking
- Personal phone directory
- Integrated voice mail with customizable outgoing messages

- Text chatboard
- Proxy support for use behind a firewall on a corporate network
- Caller ID

You can download a functional, but limited, copy from http://www.netspeak.com. The limitations of the trial version include the following:

- Three-minute talk time per call
- Only one active line instead of four
- Integrated voice mail limited to one outgoing message and two received messages

You can order an activation key online or by regular telephone at (561) 998-8700.

FreeTel

FreeTel's name tells you the most important thing to know about the program — it's FREE! That's not for a limited-use demo version or for a trial copy that expires in 30 days, but the fully operational program — free to any noncommercial user. If you're using FreeTel for business purposes, you get a 30-day free trial, after which you must pay for a license.

At this point, you must be thinking, "What's the catch?" Well, here it is: advertisements. While you use FreeTel, ad banners stream through a small box near the top of the FreeTel window. Your call is not affected in any way, nor is any of the functionality of the program, but you have no way to turn off the ads, at least not yet. In the near future, FreeTel promises to release a for-pay version of the program that enables you to turn off the promos.

FreeTel uses the now familiar server model for establishing connections to other users.

The program has the following features:

- Full-duplex sound support
- Text chat
- File transfer support
- Automatic sound compression adjustment
- Web page linking for placing calls (similar to the Internet Phone scheme)
- Caller ID with a one-line message explaining the reason for the call

You can download FreeTel from its Web site at http://www.freetel.com. You also can purchase the ad-blocking feature, or an enhanced version, FreeTel Personal Edition, from the Web site.

Voxware TeleVox

Voxware, a leading voice technology software company, provides its codecs and other technology to other companies. In fact, you can find Voxware's codecs in NetMeeting, Netscape Conference, and White Pine's version of CU-SeeMe, among other places. Besides selling technology to others, Voxware produces a few products of its own, including an Internet telephony program called TeleVox.

Features of TeleVox include the following:

- Server-based calling with the capability to create private groups (similar to the Internet Phone private topics feature)
- Multiparty conferencing, which enables as many as five people to be together on one call
- Full-duplex capability
- File-transfer capability
- Text chat
- Caller ID and call blocking
- Netscape interface, which enables you to automatically launch Netscape to look at a page featured in a window at the bottom of the TeleVox screen

The coolest attribute of the TeleVox program is its use of Voxware's voice technologies, which can provide

- High-compression ratio (good for lower bandwidth connections)
- *Voice fonts* and effects for altering the sound of your voice. For example, you can raise or lower the pitch of your voice or invoke an entirely different sound, such as a robot or a whisper.

TeleVox supports only Windows 95 and NT, so if you have Windows 3.1, don't bother trying to download this program. A new beta version of TeleVox, which is H.323 compliant, is now available on the Voxware Web site — you'll soon be able to use TeleVox to talk with other H.323 phones. TeleVox has some neat features, so keep your eye on the Voxware home page at http://www.voxware.com.

DigiPhone

Third Planet Publishing, a subsidiary of Camelot Corporation, produces DigiPhone, which you may have seen in print ads in *Wired* and other Internet-related magazines. DigiPhone does not have a downloadable demo yet, but promises to have it available soon on the Third Planet Publishing Web page, at `http://www.planeteers.com`. Two versions are available: DigiPhone, which is the telephony application, and DigiPhone Deluxe, which includes Netscape and some additional features.

DigiPhone is based on a direct-connection process, using an IP or e-mail address to find the other party. DigiPhone has a feature that enables people with dynamically assigned IP addresses to enter an IP range (which hopefully includes the address assigned to the person you want to call) and DigiPhone then searches for that person.

DigiPhone's features include the following:

- Full-duplex calling
- Adjustable compression levels
- Encryption
- Personal and DigiPhone Global directories

  Note: The Digiphone Global directory is not viewed in real time. You send a query to it (name, place, and so on), and it replies via e-mail.
- E-mail notification to the person you're calling if he or she doesn't answer
- Caller ID

The Deluxe version of DigiPhone adds conference calling, call recording, text/voice caller ID, voice messaging, and voice effects.

You can purchase DigiPhone and DigiPhone Deluxe in your local software store. Deluxe comes with a free copy of DigiPhone that you can give to a friend, so you always have someone to talk with.

Speak Freely

Here's another free program, like FreeTel. So you're probably waiting to hear what the bad news is. Well, we'll get to that first. Speak Freely likes a great deal of bandwidth, and if you have a 14.4 Kbps modem, this program probably is not going to work too well.

That said, Speak Freely is a pretty neat program, and it doesn't cost you a penny. It was produced by John Walker, the founder of Autodesk.

Speak Freely is not a commercial program, so don't expect all the support that this concept may or may not imply. Calls are placed by entering an IP address or host name directly.

The features you can expect to find include the following:

- Full-duplex capability
- Online directory of users
- Photo display capability
- Conference calling
- Broadcast and multicast capabilities
- Highly secure encryption (similar to PGPfone, a product we discuss in Chapter 7)

If you're a bit of a power user, or if you're just plain feeling adventurous, try a download of Speak Freely. To download this program, go to `http://www.fourmilab.ch/speakfree/windows/speak_freely.html` (that sure is a long one). There you find links to the download page and an excellent online manual.

Other PC programs

The preceding sections in this chapter do not provide an exhaustive listing of what's available out there on the 'Net for your Windows PC. We simply don't have room to mention all the available programs in detail here — and new programs come and go so often that we'd surely leave someone out if we tried to be exhaustive. The following table lists a few more programs that you might want to try out:

Product name	***Web site***
IBM Internet Connection Phone	http://www.ics.raleigh.ibm.com/ics/icphone.htm
Connectix Videophone	http://www.connectix.com
Netiphone	http://www.netiphone.com/
VDOPhone Internet	http://www.vdolive.com/vdostore/vdophone.html
CineVideo/Direct	http://www.cinecom.com/

Installing Internet Phone on your PC

Installing Internet Phone is simple. Just download it from VocalTec's Web site and run the installer program, as we detail in the following step-by-step instructions:

1. **Establish a connection with your ISP.**
2. **Launch your Web browser program (Netscape, Microsoft Explorer, or whatever you use).**
3. **Point it to the following URL:** `http://www.vocaltec.com/download.htm`.
4. **Click on the underlined link that corresponds with the version you can use on your computer.**

   We can't tell you exactly where to click on the page because the exact place may change as the page is updated. Trust us, you won't miss it. The link takes you to a page that lists several FTP sites from which you can download the program. (FTP, which stands for File Transfer Protocol, is a standard way for downloading files from the 'Net.)
5. **Click on the FTP site nearest you (or anyone for that matter).**
6. **Your browser automatically downloads the archive IPHONExx.EXE (xx is the build number, or version, which changes as new versions are released).**

   Depending upon which browser you are using and how you have it configured, you may have to click on OK in a dialog box to start the download.
7. **Quit Netscape and disconnect from your 'Net connection.**
8. **Double-click on the IPHONExx.EXE icon.**
9. **Click on OK in the WinZip Self-Extractor dialog box that pops up.**
10. **Internet Phone automatically runs its Install Shield (see Figure 6-1), which guides you through each step in the set-up process.**

    You don't find anything tricky here. The Install Shield simply prompts you to click on some buttons and make a few choices. (The correct choices and buttons depend on your specific computer and peripherals.)

    Congratulations, you're done!

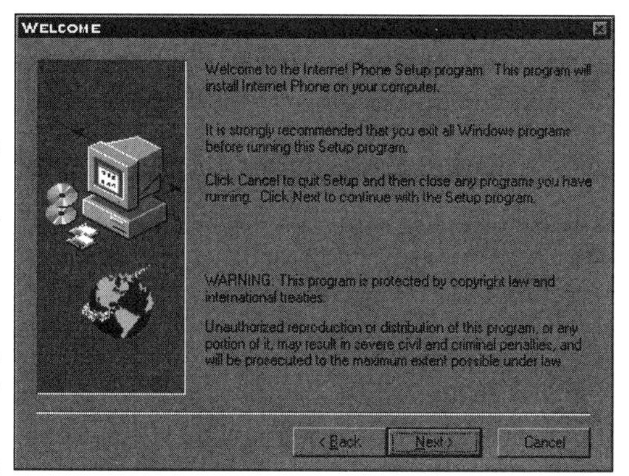

Figure 6-1: Just follow the instructions in the Install Shield — it's super easy!

Installing Other Programs Is Just As Easy

Most of the programs we mention in this chapter download in much the same way. You encounter compressed files in two flavors:

- **Self-extracting archives:** Look for an .exe at the end of the filename.
- **Zip archive files:** Look for a .zip at the end of the filename.

Both types are pretty easy to handle. As we mention previously, the self-extracting files can be decompressed via the Windows program. For the Zip archive files, you need an extra program such as WinZip. (See the sidebar "WinZip'ing along.")

You don't have too much to worry about downloading other programs. We *do* recommend that if you're installing a self-extracting archive file, you create a new folder, called TEMP or INSTALL, before you run the file. Although this step is not necessary for all software (Internet Phone doesn't need this step, for example), some .exe files do not prompt you for a location to place the files they extract, but instead dump them all in the same directory as the .exe file itself. You can end up with a real mess, especially if you download more than one file.

Put all downloaded .exe or .zip files in a special directory (a folder in Windows 95) on your hard drive before running them. This keeps things from getting messy (files strewn all over your hard drive).

WinZip'ing along

WinZip is a great program to have if you're doing a lot of downloading (or uploading) from the 'Net or from bulletin boards. It's a shareware program available from WinZip's home page (at `http://www.winzip.com`) or from lots of other sites on the 'Net. (Our favorite is clnet at `http://www.shareware.com/`.) WinZip enables you to handle .zip files like a pro. All you need to do is select the file, click an extract button, choose a directory, and leave the program alone to do its job.

The neatest thing about WinZip is that it can help you *uninstall* programs if you no longer need them. Sometimes programs hide files all over the place on you, and you can never find them all if you try to delete them manually. WinZip does it all for you. Now that's a neat trick.

Chapter 7

Getting Your Mac OS Computer Set Up for Internet Telephony

In This Chapter

- Understanding your minimum needs for Internet telephony on a Mac
- Deciding on the Mac to have for Internet telephony
- Choosing Mac programs
- Installing Mac telephony programs

We like Macs. In fact, we like them so much that almost all of the computers in our company are Macs. We've got bunches of them. So even though most of the Internet telephony programs out there are for those *other* machines, Mac users can take heart that someone is looking out for their interests (us) and we've included them in this book.

And by the way, Mac users should peruse the Windows portions of this book, if for no other reason than to be able to point to yet another way in which Windows 95 does not match up to the ease of use of the Mac OS (the Macintosh Operating System). Macs are far easier to outfit for Internet telephony and video programs because these computers have been targeted for users who, at home and at work, tend to concentrate in graphical and display-oriented applications.

So if ever a "plug and play" claim was to be made with Internet telephony or video, it is with the Mac platform.

Note: During the past year, the Mac OS computer marketplace has seen an explosion of clones. These computers (which are made by a whole bunch of companies, including Power Computing, Motorola, and UMAX) really are Mac-compatible, so if you use one of these, and it meets the basic Mac requirements we mention in this chapter, it should be just fine for Internet telephony. In fact, these Mac clones use the latest PowerPC CPUs and have all sorts of cutting-edge stuff packed into them, so they should be excellent platforms for talking or videoconferencing on the 'Net. We're not just saying that either — for all of our testing and much of our day-in, day-out use of Internet telephony programs on the Mac OS, we use a Motorola StarMax clone! So, when we say *Mac* in this chapter, we really mean Macintosh or Mac OS clone.

Mac Minimum System Requirements

Before we get into the nitty-gritty details of what your system needs to take part in this Internet revolution, make sure that you know the absolute minimum requirements, which include

- Mac LC II or higher (a Mac with a 25 MHz 68040 processor for some programs)
- System 7.1
- 8MB RAM
- MacTCP or Open Transport
- Macintosh Sound Manager 3.0

Mac Recommended System Requirements

Okay, the preceding section tells you what you need, but you may be wondering, "What do I *really* need?" Glad you asked. You really, really need at least a 68040-driven Macintosh (any Quadra or Performa 630 series) to run the programs at a reasonable speed. A Power Macintosh is better (much better, actually), so if you have an ancient Mac (such as an SE or a Mac Classic), it's time to upgrade. All of the new Macintosh and Mac OS clone computers available now have PowerPC chips.

Most recent Macs come with 16MB of RAM as the standard, but if you have only 8MB, spend the money and upgrade to 16MB.

So what's our best Mac solution? Try this:

- PowerPC or high-end 68040-powered Mac
- System 7.6
- 16MB RAM

More On What You Need to Make Your System Work

The following sections in this chapter go through each of the preceding requirements in a little more detail. If your system is deficient in one area or another, check out that particular portion of the chapter and then feel free to skip to the second half of the chapter, which gets into the specific Mac-based telephony programs available today.

68030, 040, PowerPC

Depending on which program you are using, a minimum of a 68020 or higher processor (CPU) is required. DigiPhone requires a 30 MHz clock speed 68030, whereas PGPfone requires a 68LC040 (Quadra or later).

The Golden CPU Rule still applies: Because these programs use your computer's CPU to compress and decompress the audio or video signal, the faster your processor, the faster you can perform these compressions and decompressions. DigiPhone, for example, uses the GSM protocol for its highest quality audio connection, but this runs only on PowerPC Macs.

You are best off with the fastest Mac that you can get (read that as Power Macintosh), of course, because even a fast 68040 can't support the highest quality protocols. To adapt the old adage: You can never be too rich or too thin, and you can never have too much clock speed. See Table 6-1 for a list of suggested Mac systems.

Table 6-1 Some Common Mac Systems

Model	*Processor*	*Full- or Half-Duplex*
All Power Macs	PowerPC	Full
All Centris	68030	Full
All Quadras	68040	Full
LC IIs and IIIs	68030	Half
Mac OS Clones	PowerPC	Full

Some of the most popular programs for the Macintosh require at least a 68040 processor to work at all, and at least one (Netscape Communicator) is for PowerPC-based Macs only. When we discuss each program in subsequent sections of this chapter, we make sure that you know about any special system requirements.

If you have a Macintosh Performa, you can easily tell whether or not you have a PowerPC CPU by looking at the model number. Four-digit numbers, like Performa 6116, are Power Macs, while those with three-digit numbers, like Performa 630, are powered by the older 68030 or 68040 Motorola CPUs.

8MB RAM

This is another bare-bones minimum. You'll probably find that 16MB is required, especially if you are running a Power Mac (which typically uses more RAM for the system software) or if you have a lot of extensions running in your operating system. In fact, if you have a Power Mac, you will be much happier with even more than 16MB.

Don't know how much RAM you have? From the Finder, pull down the Apple menu in the top-left corner of your screen, choose About this Macintosh (if you're using Mac OS 7.6 or later, choose About this Computer), and your Mac tells you. You also see which version of the system software you are running. (Read on to see why that's important.)

System 7.0 or greater

The latest Mac operating system level, System 7, was released several years ago and has since undergone many major improvements and releases. System 7.6 is the most recent and preferred version for Internet telephony, but if you don't have it and don't want to pay for it, you can download a free, older version (System 7.0.1) from Apple's Web site (`www.info.apple.com` or `www.support.apple.com`) or from the Apple support area on AOL (keyword: `Apple`). If you want System 7.6, you have to pay Apple for the upgrade kit, but in return you get a lot of neat new features, such as the Apple Guide help system. System 7.6 also comes with the extra system extensions that you need, such as SoundManager and Open Transport.

SoundManager 3.x

Apple SoundManager 3.0 or higher is required for all the Internet telephony programs we discuss in this chapter, but you will be best off if you install SoundManager 3.2.1 or higher. (SoundManager 3.2.1 is the latest version as we write this.) SoundManager comes with System 7.6 software and is available for free download at the Apple WWW sites that we list in the preceding section, as well as on AOL.

MacTCP

MacTCP is Apple's previous TCP/IP stack for the Macintosh. If you are using an older version of Macintosh system software on your computer (System 7.5.1 or earlier), you probably have MacTCP installed — not Open Transport. MacTCP does pretty much the same thing as Open Transport, but we prefer Open Transport for two reasons.

First, it's a lot easier to set up and use — especially if you have more than one network configuration. (For example, you may have a laptop with one setting for dialing the Internet by modem, and another for when you're back in the office and hooked up to the LAN.) Second, Open Transport is the system that Apple currently supports, and increasingly you'll find that programs require it and won't work with MacTCP.

OpenTransport can be downloaded from Apple's Web site (`http://www.apple.com`) for free, if you don't have it. Users of System 7.5.3 and later should already have it, as it's part of the system software package.

Installation is a breeze: Just drag the icon into your closed System folder and click on OK when the dialog box asks you "Do you want to put sound manager in the Extensions folder?" and your Mac will put it in the right place for you. When you restart your Mac, SoundManager will be active.

To find out which version of SoundManager you have, open your System folder and then open the Extensions folder. Click on the SoundManager icon and choose File⇨Get Info. The dialog box that pops up tells you the version. If you can't find SoundManager in the Extensions folder or if you have an earlier version, then you need to upgrade. You can download it at `http://www.apple.com`.

TCP/IP-Open Transport or MacTCP installed — Open Transport 1.1.2 preferred

Open Transport is Apple's program that allows your Macintosh to talk over a TCP/IP network. It basically does the same thing for your Macintosh that Winsock does for a PC (as we discuss in Chapter 6). Because the Internet is a TCP/IP network, your Mac will not be able to speak the "language" of the Internet unless you have Open Transport installed.

As of the writing of this book, Version 1.1.2 is the latest release of Open Transport and is preferred because it performs the best and has the fewest bugs.

Sound cards not needed

One of the neat things about Macs is that they come with many extra items right out of the box — items that you would have to go out and buy for many PCs.

One of these items is built-in sound. All Macs have some sort of built-in sound capacity, and most newer machines (such as Power Macs) have built-in 16-bit audio both in and out.

This built-in audio makes getting an Internet telephony program up and running a lot easier, because you don't need to go out and buy a sound card, install it, configure the drivers, and so on. In defense of the PC world, Windows 95 makes these installations much easier, and many of the new multimedia PCs come with built-in sound support, just like Macs did years ago.

Mac Internet Telephony Programs

When you ask people, "Should I buy a Mac or a PC?" you often get this response: "Many more programs exist for PCs, so get a PC." Depending on the application, this is sometimes true and sometimes not.

Unfortunately, in the case of Internet telephony, the advice is true. The PC programs greatly outnumber the Mac programs.

Nonetheless, the available Mac programs are powerful, and most of the PC vendors have announced their intention to launch Mac clients. So expect this disparity to change quickly as Internet telephony and video expand in use.

Netscape Conference

Like Microsoft, its big competitor in the Web browser market, Netscape has become a major player in the Internet telephony field. Also like Microsoft, Netscape bundles an Internet telephony program with its latest browser package. Unlike Microsoft, however, Netscape offers its program, Conference, on several different platforms, including — you guessed it — the Mac OS.

Conference has the following system requirements:

- Power PC computer
- System 7.5.5 (System 7.6 is strongly recommended)
- 16MB RAM
- Open Transport 1.1.2

These requirements are for the latest beta version of Conference available as we write this chapter. A final version will be out later in 1997 and could have slightly different requirements.

Conference is an H.323-compliant Internet telephony program, so it should be compatible with other products such as Microsoft NetMeeting or Intel Internet Video Phone. Unfortunately, the beta version that we tested didn't have all of the required codecs installed, so the compatibility was more theoretical than actual. By the time the final version comes out, this problem should be resolved.

Among the features of Netscape Conference are the following:

- Full-duplex audio
- Direct calling by IP address
- A Directory Location Service (or DLS) that lets you make calls by typing in a person's e-mail address or by searching a Web-based listing of online users
- Text chat
- File transfer
- A "business card" that provides information about callers, including a picture

As is the case with many Internet telephony products, unfortunately, the Mac version is a bit behind the Windows 95 version in features and development. As we write, however, Netscape has just released an alpha (in other words, not quite ready for public consumption) version of Conference for the Mac that includes most of the features of the Windows version, including a whiteboard and a collaborative Web browsing feature. Hopefully, by the time you read this, the Mac and Windows versions of Conference will be on an equal footing.

Conference is available freely over the Internet as part of Netscape Communicator (the new, improved, and expanded software program that is taking over for Netscape Navigator). You can't download Conference separately; it comes as part of the total download. You can find Communicator at `http://www.netscape.com`.

Make sure you download the "full install" package of Communicator because the minimum install version does not include Conference. We don't want you to waste your time doing a huge download twice.

CU-SeeMe

CU-SeeMe is the original Internet videoconferencing program. First developed as freeware by a team at Cornell University (the CU part of the name), a commercial version, appropriately called Enhanced CU-SeeMe, has been released by White Pine Software in New Hampshire. The commercial version costs a nominal amount, about $69.

Both versions are available for you to use: The latest freeware version can be found at `http://cu-seeme.cornell.edu`, while a 30-day demo version of Enhanced CU-SeeMe is available at `http://www.cu-seeme.com`.

CU-SeeMe is available for Macs and PCs — and the two versions are compatible, so you can conference with both Mac and PC users. CU-SeeMe enables you to connect to another person either directly, by entering an IP address, or to a multiparty conference by connecting to a reflector (server) site.

A *reflector site* is a location on the 'Net that runs a special UNIX or Windows NT CU-SeeMe program (Reflector). This program reflects signals back to those who are signaling with that site so that multiple parties can log onto the reflector and participate in conferences. You find out more about reflector sites in Chapter 13.

Because it was designed with 10-megabits-per-second LANs and campus backbones in mind, the original CU-SeeMe doesn't work very well with the low-bandwidth modems that many of us have plugged into our computers at home. The bandwidth is just not adequate. This problem isn't really too bad with the video transmission, but the audio codecs in the freeware version of CU-SeeMe aren't really designed to work at all on a modem connection.

Fortunately, that problem is being addressed by the developers of the commercial version. Enhanced CU-SeeMe offers color capabilities and is designed to run well over lower-bandwidth Internet connections, such as a 28.8 Kbps modem connection, as well as over higher bandwidths as with the original program.

If you use the Cornell University freeware version of CU-SeeMe, you can't see the video of people who are sending color video with the Enhanced version of CU-SeeMe.

Both versions of CU-SeeMe enable you to do the following:

- Participate in a video conference one-on-one.
- Team with as many as eight active participants.

  Note: More than eight participants can be on a CU-SeeMe reflector site at once, but you are limited to opening only eight video windows at a time. You can easily close some of your open windows to see others, though.
- Participate in an audio-only mode.

You can always use an Internet videoconferencing program for audio-only conferencing if you want or need to. This is particularly effective if you have a slow (dial-up) Internet connection.

With CU-SeeMe, you can also do the following:

- Watch broadcasts (one-way video).
- Share whiteboards or text chats.

  Note: A *whiteboard* on your computer is just like a whiteboard in your office or classroom. You can draw on it, put text on it, or even paste things (such as pictures) on it.
- Just *lurk* at a public reflector site (watch the video, but not send).

Features of Enhanced CU-SeeMe include the following:

- Caller ID
- Password security
- Selectable audio codecs (so that you can adjust audio compression based upon your bandwidth availability)
- Color video
- Phone book for people and reflector sites

If you have a Mac that meets our basic requirements for Internet telephony, you should be able to run the freeware version of CU-SeeMe. The Enhanced version requires a minimum of a 68030 Mac. We think you'd be a lot better off with an even faster Mac, especially if you want to keep multiple video windows open at once.

CU-SeeMe is not yet H.323 compatible, so you won't be able to communicate with anyone using a different program.

You find out more about using CU-SeeMe in Chapter 13.

Internet Phone for Macintosh

Internet Phone, by VocalTec, was one of the original Internet telephony programs available on the 'Net, and it's still one of the most popular. First available only for Windows computers, you can now use Internet Phone on your Mac. In fact two very similar versions of Internet Phone for the Mac are available on the VocalTec Web site (`http://www.vocaltec.com`) — a PowerPC-only version and a version for Macs with 68040 CPUs. Please note that if you have an older 68030 or earlier Mac, you are out of luck — we warned you!

Internet Phone allows you to make phone calls to other Internet Phone users (Mac or Windows — but not users of other products because Internet Phone is not H.323 compatible) by directly entering an IP address or by entering a chat room in VocalTec's Global OnLine Directory, or GOLD — a network of servers that lets you find people who are online using Internet Phone. Literally hundreds of chat rooms are on the GOLD, with subjects ranging from adults-only to Hong Kong, so you should have no trouble finding someone with whom you can chat. You can also create your own chat rooms — including private ones that only you and your friends know about.

Internet Phone's features include:

- Full-duplex audio
- A phone book to store frequently called names
- An animated assistant that uses a cartoon figure to let you know what's happening with your call
- A "do not disturb" feature, for those times when you only want to initiate, and not receive, calls
- Selectable codecs, to fine-tune your audio quality

We're sorry to once again have to tell you about some features that Windows users get, but Mac users have to do without. These include:

- Videoconferencing capabilities (sorry, Win95 users only)
- File transfer
- Text chat
- Whiteboard

We talk about using Internet Phone in more detail in Chapter 10.

PGPfone: say what you please

This Internet phone program was developed by Will Smith and Phil Zimmerman of Pretty Good Privacy (PGP) fame. You may have heard of Phil and the big brouhaha over the PGP encryption program in the news, in *Wired*, on the Web, or somewhere where people like to talk about the Internet (which is just about everywhere nowadays).

PGP is a program that allows you to send encrypted files, e-mail, and so on, to your friends, colleagues, and others with whom you share your public encryption key. Prying eyes can't read what you are saying, but your intended reader can. It was made available over the Internet, and some regulators got upset because versions of PGP were being taken off the Internet by people outside of the country, possibly violating technology export laws (turns out that encryption is considered a national security issue by some people).

What does this have to do with Internet telephony? Well, PGPfone uses a similar encryption scheme to make your phone calls secure. No one besides you and the other party "on the line" can hear what you're saying.

Having said all that, PGPfone is a pretty good program (no pun intended). In addition to its unique security functions, it offers the capability to make secure calls modem-to-modem or over an AppleTalk network, in case you need to call someone who is not on the Internet (though why would you want to talk to one of *those* people?).

With PGPfone, you place Internet phone calls directly by entering the IP address of the person you are calling.

Features of PGPfone include

- Support for full-duplex sound
- Two-way, secure file transfer
- A phone book for storing phone numbers or Internet addresses of those people you call most frequently
- Adjustment of both the encryption and audio compression standards to meet bandwidth and processor limitations

When we wrote the first edition of *Internet Telephony For Dummies*, PGPfone was an early beta version, and it was free. That's ended — sort of. The last version of the beta edition of PGPfone (for both the Mac and Windows 95) is still available, but a new commercial version has been released for the Macintosh only. You can find out about both the new version and the old beta versions at the PGP Web site (`http://www.pgp.com`); however, PGP offers technical support for only the commercial version.

We spent some time talking with Will using an early beta version of PGPfone, and the sound quality was great. Keep your eye on this one.

Because of its encryption and the legal battles that PGP has had to deal with, PGPfone is available only in the U.S. and Canada. To prove that this is going to be used just in the U.S. or Canada, you have to go to the MIT server that is handling distribution (`http://web.mit.edu/network/pgpfone/`) and fill out a form stating your location and citizenship/residency. You can then download the program, install it, and start talking securely.

If you are still thinking that you can download the program from outside the country, think again. The PGPfone server uses something called *inverse mapping* to query back over the Internet and find out whether you really are located in the U.S. or Canada. This may not work even though you are in the U.S. or Canada, if your ISP doesn't support this function. Don't get frustrated, because you can download PGPfone from several other, unofficial sites. These sites may change, but if you go to `http://www.yahoo.com` or `http://www.lycos.com`, you can search for them. You may not satisfy your need for immediate gratification, but you can probably find the software you want.

An old favorite with a new name: DigiPhone for the Mac

One of the first, and most popular, Mac 'Net telephony programs was one known as NetPhone. The company that originally developed and sold NetPhone (the Electric Magic Co.) sold it to another company, which renamed it EPhone and then sold it to Camelot Corp., the company that produces the DigiPhone and DigiPhone Deluxe programs for Windows computers. Camelot, in turn, renamed the program DigiPhone for the Mac and did a whole bunch of work on it, adding new features and capabilities.

DigiPhone allows you to place calls in two ways:

- Directly, by entering an IP address or host name. (Don't remember what an IP address is? Turn to Chapter 1 and refresh your memory — you didn't skip right to this chapter, did you?)
- Indirectly, through a server known as a NetPub

Logging onto a NetPub enables you to find people who are interested in chatting and, significantly, helps you locate people who have a dial-up ISP account with a dynamically assigned IP address.

Other features of DigiPhone include

- An address book
- Full-duplex sound on Macs that support it
- Encryption for your calls — to ensure that no one is listening
- A variety of audio codec options (see Chapter 2 to find out what codecs are) that allow you to optimize DigiPhone for your Mac's hardware and Internet connection limitations
- Support for a protocol called VAT

Note: VAT is an Internet voice protocol used by many of the early UNIX 'Net conferencing programs. Many UNIX computers use VAT for Internet phoning, so you can use DigiPhone to call anyone with a UNIX machine running VAT.

You can download a 30-day trial version of DigiPhone from the DigiPhone Web page at `http://www.digiphone.com`.

Other products for the Mac

Although the list of Internet Telephony products for Mac OS computers is much shorter than that for Windows 95 machines, we have a few more options for you to check out:

- **Connectix Videophone:** This videoconferencing product from the makers of the QuickCam comes in both Mac and Windows flavors. You can download a trial version and find more information at Connectix's Web site: `http://www.connectix.com`.
- **ClearPhone:** This is a Mac-only audio- and videoconferencing product. Trial version downloads and more details can be found at `http://www.clearphone.com`.
- **Maven:** Maven is an older, audio-only, free conferencing program that can be used to talk with other Maven users, users of the Cornell version of CU-SeeMe, and UNIX computers using the VAT protocol. You can find Maven at various download sites on the 'Net — it doesn't have a home page of its own. Use a search engine such as AltaVista (`http://www.altavista.digital.com`) to find it, or search on your favorite Mac software site.

As you have no doubt seen by now, Mac versions of Internet telephony software tend to lag behind their Windows counterparts. However, we think you'll see more Mac versions of some of the popular programs pop up in the not so distant future. For example, Microsoft is working with some other companies to produce a Mac version of NetMeeting — something we'd love to see! So take a moment to browse through the previous chapter even if you have no intention of ever buying a PC — you may find some future Mac programs listed there.

Installing Internet Phone on Your Mac: Easy as (Apple) Pie

To give you an idea of what is involved in installing a program on your Macintosh, we provide the following example for Internet Phone. Honestly, we had to stretch to make installing Internet Phone sound at all like a multistep process. It is really easy. Here's what you do:

1. **Make sure that you have satisfied all the hardware and software requirements that we list earlier in this chapter.**
2. **Make sure that you are all set up with your ISP.**

   If you need to brush up on the details for completing this step, refer to Chapter 3.
3. **Log onto the Internet (through your ISP).**
4. **Double-click on your Netscape (or Internet Explorer or other browser) icon.**
5. **Enter** `http://www.vocaltec.com/download.htm` **in the URL window and press Return.**
6. **Click on the link to Internet Phone for Macintosh.**

   This link takes you to the window shown in Figure 7-1.
7. **Click on one of the links labeled "download."**

   If the site you choose is busy, simply click another download link on this page.

   Your browser automatically puts the file in the location you use for storing downloaded files. If necessary, you can change this setting in your browser's Preferences menu.
8. **Decompress the downloaded file.**

   Use Aladdin StuffIt Expander or whatever decompression program you prefer. (See the "StuffIt? You StuffIt!" sidebar.) If you have your Web browser set up properly, decompression is automatic. When you decompress the downloaded file, it expands into a folder called Internet Phone — which you can store wherever you'd like on your hard drive.
9. **Open the folder named Internet Phone, double-click on the Internet Phone icon, and you are ready to go.**

We didn't even get to 10 steps. Why can't all programs be so easy to install?

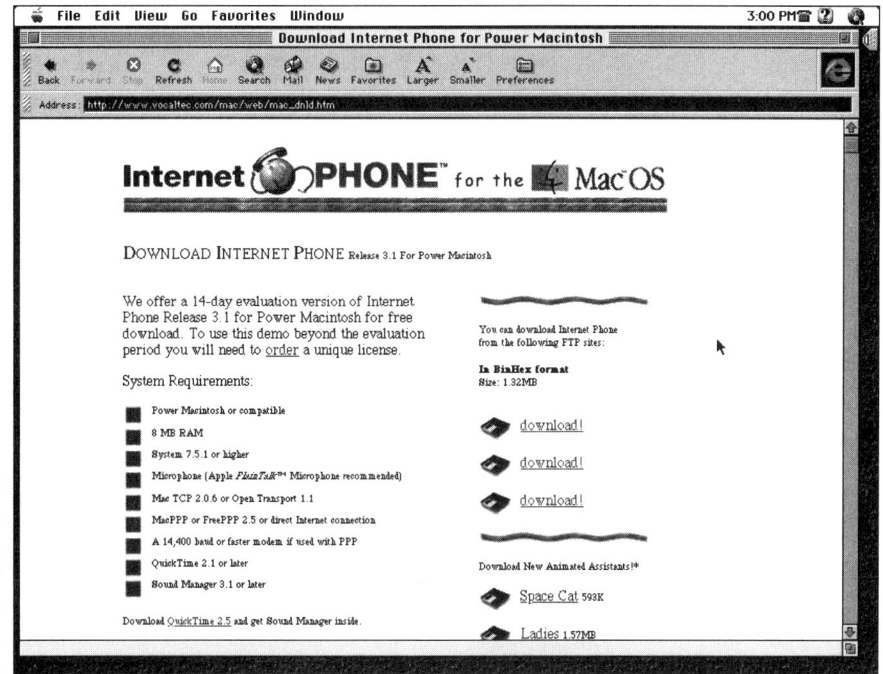

Figure 7-1: Downloading a copy of Internet Phone.

The installation procedure is about the same for all the Mac programs we mention in this chapter. Go to the Web site, download the file to your hard drive, decompress it, and you're ready to go. A few of these programs (specifically Netscape Conference and Enhanced CU-SeeMe) put files in different places on your Mac (including the System folder), so they have an installer program to do this for you. If you run into one of these installers, just follow the directions on your screen — it's really very simple.

We recommend that you download all of the programs we mention in this chapter and try each one. The user interfaces differ significantly, and you will certainly have a personal favorite in terms of look and feel.

We're the kings of "No Manual Reading for Me." But we must recommend that you read the manuals and the ReadMe files that come with your downloaded programs before you do anything else. This is really critical with Internet telephony and video programs because they change so much. Sure, it's fun just to jump right in and start using the program, but spending a few minutes reading now may save you a lot of pulled hair later.

StuffIt? You StuffIt!

If you start downloading programs from the Internet, you quickly find out that you need a program to extract the files you get. Some of the file names you see have an `.sea` on the end (self-extracting archive) whereas others have an `.sit` (StuffIt); you may also see file names that end with a `.bin` (Mac binary) or an `.hqx` (binhex). Even worse, you may get a file with a `.bin` or an `.hqx` and then find an `.sea` or `.sit` inside. Sound confusing? Don't worry, because you can make this whole process transparent. Download the Aladdin StuffIt Expander at `http://www.aladdinsys.com`, and you'll be all set. It's free, and all you have to do to use it is drag the file over the Expander icon. In fact, if you use Netscape Navigator as your Web browser, you probably already have StuffIt Expander on your computer. (It's part of the download package from Netscape.)

You can even make your Web browser do it automatically. If you are using Netscape, just choose Options⇨Preferences, click on Helper Applications, and select StuffIt Expander for all of the file types we mention in this sidebar. Then, the next time you download a file, Netscape automatically decodes and expands it for you!

Chapter 8

The "Now Ready for Real-Time" Audio and Video Players

In This Chapter

- Couch potatoes on the 'Net
- Getting hooked into streaming audio and video
- Watching the tube

The Internet telephony and videoconferencing programs that we talk about in the previous few chapters are all interactive in the sense that you can use them to interact with one or several other people on the 'Net. The real-time audio and video players that we discuss in this chapter are for those times when you don't want or need to talk back.

Until recently, the Web has been predominantly passive, with Web pages offering you text and graphics. Maybe you could download an audio file or a QuickTime movie from a Web site somewhere and then play it back on your computer later.

In the last year, however, live media has exploded on the 'Net. You can use the programs we discuss in this chapter to listen to the radio, get news updates, sample music, watch TV broadcasts, and lots more.

We mention the programs we discuss in this chapter for completeness and so that you do not confuse them with true Internet telephony and video programs. You may want to have these streaming audio and video players on your machine as well, so consider this chapter an extra on our part.

By the way, if you think that the telephony and videoconferencing programs are worrying the telephone companies (and they are!), you can bet that these applications are giving the broadcast industry something to worry about. (We've been noticing people a little on edge lately.)

How Live Audio and Video Products Work

Unlike the "download and play back later" media of the recent past, *streaming* audio and video products play back media in real time. Rather than wait for the file to download, these programs decode the data as it streams over the 'Net onto your machine and begin playing it back immediately. The data file (which can be quite large, especially in the case of video) never gets placed on your hard drive (unless you want it there), so you don't need to worry about disk space or ending up with a bunch of extra files on your desktop. You just sit back and listen or watch.

The player applications are integrated into your Web browser, so all you need to do is go to the Web site that has the content you're interested in, click what you want to see or hear, and have your computer do the rest.

In Chapter 19, we list some interesting places for you to visit, and we tell you where to go to find more.

Is it live, or is it on demand?

Streaming audio or video content can be broken down into two categories, live and on demand, as follows:

- **Live audio or video:** This category is just what you think it is — live speech, music, or video, encoded on the fly and sent downstream to you, the listener or viewer, as it happens. Not all streaming audio and video programs support this real-time encoding, although all support the real-time decoding on the receiving end.
- **On demand:** In this type of streaming, content must be recorded and encoded ahead of time by the content provider and then placed on its Web server for your listening and viewing pleasure. The on-demand part comes into play for the receiver, because you can access the material whenever you want. This means that you can play it back, pause it, rewind it, or even fast forward it to the end — sort of like watching a videotape or listening to a CD.

As a consumer, you use virtually identical processes for accessing both live and on-demand content. Of course, the on-demand streams can't provide you with live broadcasts of speeches or basketball games, if that's what you're looking for.

Hardware? Chances are, you've got it

If you already have a sound card in your PC (none needed if you're a Mac user), an Internet connection, and a fast computer, you have everything you need.

If not, here are the details of what you need:

- Sound playback device, such as a sound card (built into Macintosh)
- Fast processor — 486 or higher for PC, 68040 for Mac
- High-speed Internet connection

Notice that we say a high-speed Internet connection. With a 14.4 Kbps modem connection, you can get AM-quality audio playback and low-quality video (1 fps at best). Higher-speed connections get you better audio quality and noticeably better video results. If you're lucky enough to have an ISDN connection, you can get a stream with full-motion video (15–30 fps if network traffic isn't too high).

What Are Your Options?

Several competing audio-on-demand and live audio technologies exist, all of which are vying to become the standard for the future.

In the audio-only realm, the current market leader is the Progressive Network RealAudio technology, with the Macromedia Shockwave, Netscape Media Player, VocalTec IWave, DSP Group True Speech, and VoxWare ToolVox technologies all following closely on RealAudio's tail.

The list of products that offer streaming video broadcasts is growing fast, with several major players competing for your viewing time. Progressive Network's RealVideo, Vivo Software's VivoActive, Xing Technology's StreamWorks, and VDOLive by VDOnet Corporation are all extremely popular options — and Microsoft's new NetShow program is competing with all of them.

Low, low price

Here's some good news: The players for these systems cost you next to nothing. In fact, most of the players are free, at least for the time being. The *client* or *playback* software is freely distributed, to make accessing the content cheap and easy for potential listeners. The licenses for the broadcast servers can be quite expensive (some are in the thousands of dollars), and this is where the companies make their money.

Plug in to the future

Helpers (software packages that help your browser do things it can't do on its own) that meet the Netscape Navigator or Internet Explorer plug-in specifications use something called an *API*, or *A*pplication *P*rogram *I*nterface, that fully integrates the helpers into your browser. Rather than run as a separate standalone program, as helper applications such as WinZip do, the plug-in code actually becomes part of the browser program and provides increased functionality.

Plug-ins enable the Web page content to be more invisible to the user because the author of the Web page can control more fully how and when the viewer can access the content. For example, a Web page can automate the playback of the embedded audio or video content so that the user can pretty much sit back and listen (or watch).

Plug-in versus helper

Most Web browsers use helper applications to perform various auxiliary functions, such as displaying graphics or decompressing files. You usually assign these functions in the browser's Preferences menu. Streaming audio and video players typically work in this fashion. If a streaming type file is selected, the browser automatically starts the designated helper application (in this case, the player), and playback begins.

Since Netscape Navigator 2.0 came out, however, modern browsers use a more sophisticated method of integrating helpers into the browser; this method uses software called *plug-ins*. Plug-in software adds its functionality directly to the browser — no separate program needs to open and do its thing. (See the sidebar "Plug in to the future.") Most browsers, such as Microsoft Internet Explorer and Netscape Navigator, support plug-ins, and those that don't now most likely will in the future. (We think you really should be using one of these plug-in capable browsers — most of the remaining alternatives just don't keep up with the demands of state-of-the-art Web pages.) Most of the streaming applications are beginning to offer components to support the plug-in standard.

Most of these streaming audio and video programs use an installer that places the various components of the program in the right places on your machine. If you're using Netscape Navigator 2.0 or later, Internet Explorer 3.0 or later, or another plug-in–capable browser, make sure that you click on `Yes` if the installer asks you whether to install plug-ins.

ActiveX?

If you follow the latest goings-on on the Web, you've probably heard of Microsoft's ActiveX technology. Like Java — its biggest competitor, and in fact its apparent mortal enemy if you were to believe some of the reports in the press — ActiveX allows small programs, called ActiveX controls (they're called applets in Java), to download from the 'Net to your computer and run. This gives content creators on the Web much greater abilities to do neat stuff on their sites — an ActiveX control can do almost anything on your PC that a regular program can do.

A small, but growing, number of companies that make streaming multimedia programs are adding ActiveX controls to their current repertoire of stand-alone players and plug-ins. If you use Windows 95 or NT and Internet Explorer 3.0 or later, you can try NetShow, VivoActive, and Shockwave (all discussed in a moment) as ActiveX controls. Have fun!

Multicasting — the future of streaming?

Currently, some severe limitations hamper the use of streaming audio and video products over the Internet. Most streaming products send an individual stream of data to each and every client program on the 'Net that requests the stream. (In other words, a Web page with a link offering streaming audio or video must send a separate stream to each 'Net user who clicks on that link.) This technique, known as Unicast, works fine up to a certain point, but as the number of people requesting the stream increases, the huge number of individual streams may overload the bandwidth availability and capacity of the routers on the 'Net.

The answer, in most expert's opinions, is to utilize a technique called Multicasting. When streams are sent via Multicast, a single stream is sent over the 'Net to all of the clients who ask for it — and only to those clients. Part of the Internet, called the MBONE (or Multicast Backbone), is already set up for this technique. The MBONE has been in operation for several years, transmitting such things as conferences and a famous Rolling Stones concert (hailed as the first concert to be broadcast via the 'Net). Unfortunately, most of the 'Net is not part of the MBONE, and the part that is typically doesn't include the dial-up ISPs that most of us use.

A group of streaming multimedia companies — including most of the companies we discuss in the following sections of this chapter — have recently formed an association to promote the development of Multicast capabilities on the Internet and to produce client software that will let you receive Multicast broadcasts. For now, only people on the MBONE — or on Multicast-capable intranets — can receive Multicast, but we hope that in the near future we all will. The efficiencies that Multicast offers will make streaming multimedia much more like the Internet TV replacement that many hope it will become.

Streaming Video Applications

To the user, all the applications we discuss here are basically similar because they are pretty much invisible — especially if they're used as plug-ins. Go to the site with the content that you want to see, click on the link, and the player starts and — voilà! — you're seeing streaming video and hearing streaming audio.

Progressive Networks' RealPlayer

Probably the most widely used streaming media system is Progressive Networks' Real Player. Chances are you know of, or even have on your computer, its predecessor, the RealAudio player. In fact, RealAudio was so successful that it was almost a de-facto standard for Internet "radio." Literally hundreds — maybe thousands — of Web sites, big and small, now offer some sort of RealAudio content. Progressive wasn't content to stick with audio only, so it recently added RealVideo to the arsenal and now distributes the RealPlayer.

You can still use the RealPlayer to play audio from one of the thousands of RealAudio-enhanced Web pages, and you can use it on a growing number of pages that incorporate RealVideo.

Two versions of the RealPlayer exist:

- RealPlayer: a free version
- RealPlayer Plus: a commercial version with extra features

Specifically, the RealPlayer Plus gives you a faster download procedure and buffers that allow you to get better quality over modem connections, as well as a scan feature that lets you search the Web from the player to find content that you like.

We recommend that you start off with the free version — see how often you use it, and determine whether the additional features of the RealPlayer Plus are worth about $30 to you.

PC versions of both RealPlayers require Windows 95, NT, or 3.1, and a minimum of a 486 33MHz processor. The Mac version is for Power Macs only. You can find the RealPlayer, an online order form for the RealPlayer Plus, plus a huge listing of sites using RealAudio or RealVideo at `http://www.real.com/`.

Microsoft NetShow

In the last year or so, Microsoft has decided that its future is the Internet. Thanks to its vast resources of money and talent, it's gone from 0 to 60 in record time. For example, Internet Explorer went from also-ran to heavy-weight contender in almost no time. For Internet telephony, Microsoft offers NetMeeting, which came to market late compared to its competition, but now sets the standard in many ways (including literally, as it is the first H.323 and T.120 application available). And Microsoft didn't forget about streaming video applications, either. Its foray into this field is called NetShow.

One of the most exciting features of NetShow — at least to some users — is that it incorporates Multicast capability for use on networks that can take advantage of that feature. The NetShow client software is currently available for only 486 66MHz or faster PCs running Windows 95 or NT, but versions are forthcoming for the following operating systems:

- Macintosh
- Windows 3.1
- Various UNIX platforms

NetShow is rather new to this field, so you won't find vast numbers of sites offering NetShow video or audio — yet — but we expect the number will grow quickly. You can find a download of the player (still in beta version as we write) and more information about the program at `http://www.microsoft.com/netshow/`.

Vivo Software's VivoActive

A fast growing player in the streaming video field is Vivo Software, with its VivoActive player. VivoActive uses standard compression codecs (like H.263 and G.723 from the H.323 telephony standard) to compress and then stream Video for Windows (AVI) or QuickTime files over the 'Net. The VivoActive player is available for the following operating systems:

- Windows 95
- Windows NT
- Windows 3.1
- Macintosh

The Windows versions require a PC with a 486 66MHz or faster processor, while the Mac version requires a PowerPC CPU. You can find more information, including a listing of content providers and a free download, from Vivo's Web site at `http://www.vivo.com/`.

Xing StreamWorks

StreamWorks is a live and on-demand audio and video player that uses the MPEG standard for encoding audio and video streams for transmission over the 'Net. You can download the player for free from the Xing Web site at `http://www.xingtech.com`, although you must pay a nominal amount (about $29) to get any technical support from Xing. Figure 8-1 shows a StreamWorks playback of an *X-Files* clip.

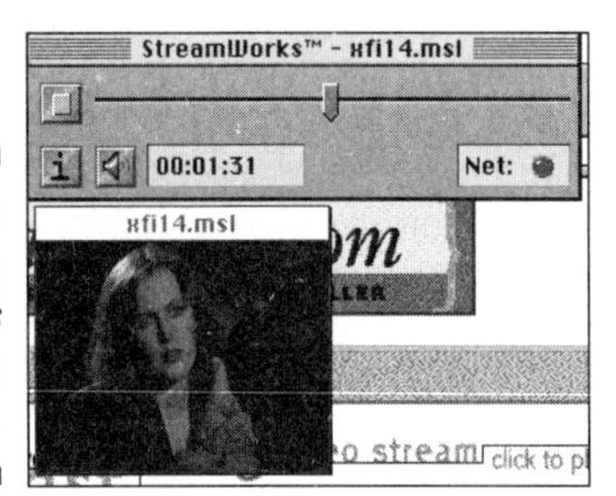

Figure 8-1: Watching an *X-Files* clip via StreamWorks.

You can use the player as either an audio-only system or as a combined video and audio player, depending on what the site that you're viewing is broadcasting. Many of the sites now available are audio only, so StreamWorks also competes with the audio players that we mention later in this chapter. Providers of StreamWorks content can offer more than one stream, so if you have an ISDN or other high-speed connection, you can choose a higher-quality connection. The player is available for the following platforms:

- Windows 3.11, 95, or NT
- Macintosh and Power Macintosh
- UNIX machines running Xwindows (if you have one of these, you probably don't need to be reading a *...For Dummies* book — but you're welcome to anyway)

VDOLive

VDOLive is a video-on-demand player that is one of the main competitors to RealVideo and StreamWorks.

VDOLive uses a new kind of compression technique called *wavelets*. Wavelets are one of the hot technologies in the compression world because they can be extremely efficient (that is, signals can be compressed by a whole lot) without losing too much quality.

VDOLive, like almost all of these programs, is a free download. You can find it at `http://www.vdolive.com`. Players are available for the following platforms:

- Windows 3.11 and 95
- Power Macintosh

VDOLive is available as both a helper application and a browser plug-in. As in StreamWorks, you can select higher-quality video streams from some sites if you have a high-speed Internet connection.

Audio-Only Apps

As is true of the preceding video applications, the audio-only players we describe in the following sections function pretty much automatically. After they're installed and your browser knows to recognize them, they work on full automatic.

DSP Group True Speech

The DSP Group's True Speech is an audio on-demand-only player that uses DSP's proprietary True Speech algorithm. This extremely efficient (high-compression) system has been licensed to Microsoft for inclusion in the Windows 95 operating system's Sound Recorder. The player is available for free download at DSP's Web site at `http://www.dspg.com`.

Supported platforms include the following:

- Windows 3.11, 95, and NT
- Macintosh and Power Macintosh using System 7.0 or greater

Macromedia Shockwave

Macromedia's Shockwave plug-in for Netscape and Internet Explorer is widely known and used for displaying cool graphics, animations, and even games in Web pages. In addition to these functions, Shockwave can be used for high-quality audio streaming from Web pages. The Shockwave player is available for the following platforms:

- Windows 3.1, 95, and NT
- Macintosh and Power Macintosh

Macromedia has several different programs for Web site authors to use while creating content — so several different flavors of Shockwave plug-ins are available. Unless you know that you want to view only one particular kind of Shockwave content, your best bet is to download the "Shockwave — The Works" package from Macromedia's Web site (`http://www.macromedia.com/`). That way, you'll be covered no matter where your Web browsing takes you!

Netscape Media Player

Netscape has developed its own plug-in streaming audio system for the Netscape Navigator Web browser: the Media Player. The Media Player uses the new Realtime Streaming Protocol (RTSP) to deliver audio streams across the 'Net or a corporate intranet and uses many of the same audio codecs (like Voxware's efficient codecs) to squeeze the audio signal down to size. Additionally, Media Player can provide *synchronized multimedia*, which lets Web content providers automatically guide your browser to different Web pages while playing back a media stream.

The Media Player plug-in is available for all of the same systems as the Navigator Web browser, including the following platforms:

- Windows 3.1, 95, and NT
- Power Macintosh
- A whole bunch of UNIX platforms

The Media Player plug-in is available for free download from the Netscape Web site, `http://www.netscape.com/`.

VocalTec IWave

VocalTec, the maker of Internet Phone (see Chapter 6), uses a *codec* (*enco*ding and *dec*oding system) similar to Internet Phone's for this streaming audio product. VocalTec offers the software for free, but you must download it in a bundled package with the demo version of Internet Phone (not really a disadvantage if you want the phone program, but it does make for a long download). You can get the IWave/Internet Phone download at `http://www.vocaltec.com`.

Supported platforms include the following:

- Windows 3.11, 95, and NT
- Power Macintosh

VoxWare ToolVox for the Web

VoxWare ToolVox for the Web uses an extremely efficient encoding scheme (similar to the one VoxWare uses in its TeleVox phone program and sells to other companies like Netscape for use in theirs — see Chapter 6) to provide excellent voice quality at low-bandwidth connections, such as a 14.4 Kbps modem connection. The on-demand-only player, available as both a helper application and a browser plug-in, can be downloaded for free from VoxWare's Web site at `http://www.voxware.com`.

Supported platforms include the following:

- Windows 3.11 and Windows 95
- Power Macintosh

Where to Start?

You should probably start by installing the free RealPlayer on your machine, as this is the program you're likely to run into the most often while online. If you have room on your hard drive and have time to spend downloading, you may want to install all these programs. Content determines which applications you need — you want to have the players that work on sites that you like to visit.

Part II

You Make the Call: Using Internet Telephony Products

"IT'S JUST UNTIL WE GET BACK UP ON THE INTERNET."

In this part . . .

It's time to get up and talking on the 'Net. In this part, we look in detail at five of the major software programs available for audio and videoconferencing over the Internet, and we get you started using them.

Each chapter covers a single product and shows you how to start, configure, and personalize it, as well as how to start making (and receiving) calls. We also show you some of the neat features that Internet telephony programs include — features you'd pay a fortune for from your local telephone company.

Chapter 9

Using Microsoft NetMeeting

In This Chapter

- Welcoming Microsoft to the fray
- Getting NetMeeting set up the way you want it
- Getting connected
- Getting collaborative

When we wrote the first edition of *Internet Telephony For Dummies,* less than a year ago, Microsoft NetMeeting was a brand-new program, still in its early beta form. We actually "stopped the presses" to include some information about NetMeeting in that edition of the book because we figured that Microsoft getting involved in Internet telephony was a big deal. Turns out we were right (though we don't claim to be geniuses just because we made this prediction — whenever the folks at Microsoft get into a new field, they do so in a big way!). NetMeeting is now one of the leaders in the Internet telephony market — a free, H.323- and T.120-compatible program, with video, audio, and lots of data conferencing stuff thrown into the mix.

NetMeeting is a full-featured Windows 95 and NT (and only Windows 95/NT at this point in time) conferencing program that is freely available from the Microsoft Web site (`http://www.microsoft.com/netmeeting`) as a part of the Internet Explorer family of products. Like Netscape's Conference (which we discuss in Chapter 12), NetMeeting is part of a movement to include Internet telephony as a component of a larger package — in this case, Web client software.

Along with all the features that are becoming standard with other telephony programs — such as text chat and whiteboarding — NetMeeting enables you to share applications between computers in a conference. For example, you can open a Microsoft Excel spreadsheet on your computer, and an image of that spreadsheet appears on the other party's screen.

Like VocalTec's Internet Phone (which we discuss in Chapter 10, in case you're skipping around), NetMeeting enables you to connect your computer to several other computers at once — although, again like Internet Phone, you can conduct a voice conversation with only one person at a time.

When making calls, NetMeeting uses a directory service called ILS (Internet Locator Service) to look up IP addresses of people. After starting up, NetMeeting immediately contacts this ILS server and lets it know that you are logged on to the 'Net and running NetMeeting. NetMeeting also gives the ILS server updated information about you — for example, your name, e-mail address, and, most importantly, your current IP address.

NetMeeting handles the ILS servers a bit differently than most of the other programs that we discuss. (By the way, the ILS servers are a new and improved version of — and the replacement for — the ULS servers that we mention a few places in the book.) Most programs use a Web browser page to display the listing of users logged into a directory server. NetMeeting, on the other hand, has a built-in display of the ILS. You can use a Web browser to view the ILS, if you wish, but the internal directory viewer works so well for us that we never do.

What Does NetMeeting Look Like?

NetMeeting has a single-window interface, shown in Figure 9-1. If your screen doesn't show everything that you see in the figure, open the View menu and make sure that Toolbar is selected. (Look for a small check mark next to this option in the View menu.) If you use Internet Explorer, NetMeeting should look very familiar — many of the buttons and controls are identical. Across the top of the window, NetMeeting offers seven pull-down menus (from left to right):

- **Call:** Places and ends calls.
- **Edit:** Provides standard editing tools, such as Copy, Cut, Paste, and Find.
- **View:** Controls what the program displays in the NetMeeting window.
- **Go:** Contains commands for opening the Web directory and other Internet Explorer programs, such as the e-mail program.
- **Tools:** Enables you to use the multimedia features, such as application sharing and whiteboarding, and to configure NetMeeting. Also allows you to switch audio and video between different people while in a multiparty conference. (You can only see and hear one person at a time.)
- **SpeedDial:** Stores the names of people you call often, like an address book.
- **Help:** Provides access to the help function.

Beneath the menu bar, NetMeeting has a toolbar with seven buttons (from left to right):

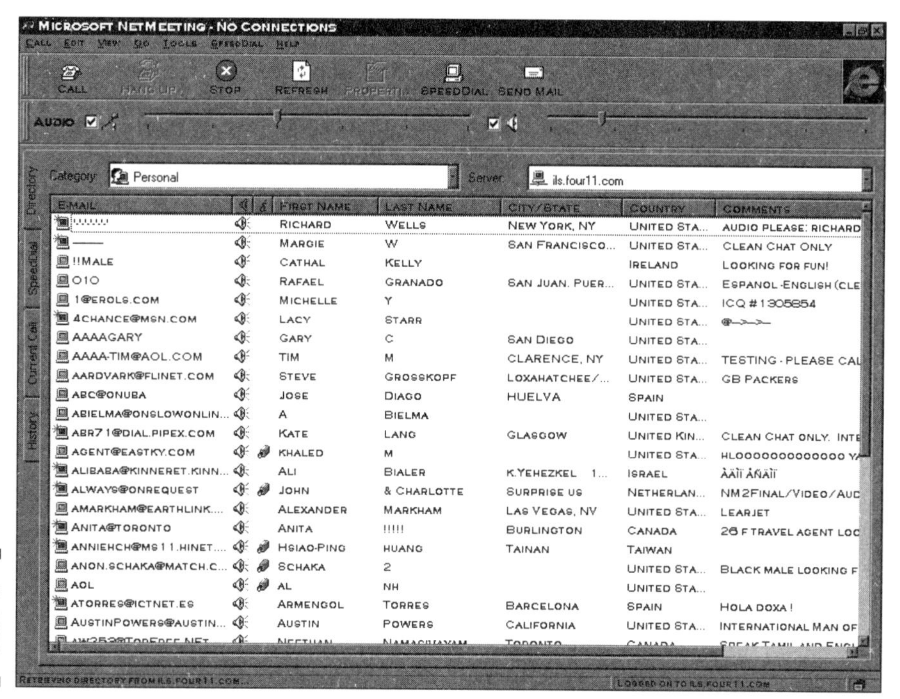

Figure 9-1: Presenting NetMeeting!

- **Call:** Click this button to place a call.
- **Hang Up:** You can probably guess what this one does!
- **Stop:** Like the Stop button on a Web browser, clicking this button stops certain actions of NetMeeting — such as loading a directory.
- **Refresh:** Another browser-like button (we told you you'd see some); you click this one to update listings in NetMeeting.
- **Properties:** This button lets you get detailed information about another user on the directory.
- **SpeedDial:** Click this button to open the SpeedDial dialog box for adding people to your SpeedDial list.
- **Send Mail:** This button opens the Internet Explorer e-mail program to send mail to a person selected in one of the NetMeeting user listings.

Before we go any further, we need to tell you something about the toolbar. If you look at the picture of NetMeeting in Figure 9-1 (or on your own computer screen), you notice four tabs running down the left-hand side of the window. The toolbar is context-sensitive — that is, its contents change depending on which tab you select.

The preceding listing of toolbar buttons applies to the Directory tab, which is where you'll probably spend a lot of your time while using NetMeeting. As you click on different tabs, some of the buttons in the toolbar change. For example, you may see some of the following buttons:

- **Delete and Delete All:** Lets you remove a selected user, or all users, from a NetMeeting listing of users.
- **Switch:** Lets you change whom you share video and audio with in a multiparty conference.
- **Share:** Turns on the application-sharing feature.
- **Collaborate:** Enables other people to edit a document in a shared application.
- **Chat:** Opens the Text Chat window.
- **Whiteboard:** Opens the whiteboard for sharing drawings, data, and pictures.

Directly beneath the toolbar, you find the audio toolbar, which has the following controls (from left to right):

- **Microphone check box:** Indicates whether your microphone is activated.
- **Microphone level slider:** Lets you increase or decrease your microphone level.
- **Speaker check box:** Turns on your speakers or headphones so you can hear incoming audio.
- **Speaker level slider:** Lets you decrease or increase the volume level on your speaker or headphones.

Like in Internet Explorer, you can move these two toolbars around your screen. Move your mouse pointer over the left-hand side of the toolbar (near the two vertical lines). After your mouse pointer turns into a little white hand, you can click and drag the toolbar somewhere else — for example, you can put the two toolbars next to each other instead of taking up screen real estate by stacking one above the other.

Directly below the audio controls are the four listing views. By clicking on the tab for a particular listing view, you change the information that the program lists in the NetMeeting window. You use all of these tabs at one time or another — some more than others. Here's the lowdown on each one:

- **Directory:** This listing shows everyone who is connected to the ILS directory server you are currently viewing (except for people who have logged on without listing their names). A Category drop-down list helps you narrow the listing, and the Server pull-down list allows you to view the user listing for a different ILS server.

- **SpeedDial:** This listing shows everyone in your SpeedDial list. In addition to listing their names, NetMeeting actually goes out to their ILS server and finds out whether they are online. Pretty handy, if you ask us.
- **Current Call:** This listing identifies all users in your current call and shows their status (for example, whether they are sending audio or video or sharing an application). This view also contains your My Video window (which displays your outgoing video) and a Remote Video window (which displays the incoming video from the other party in your call).
- **History:** This listing provides a history log of all past calls.

You can sort the information in each of these listings by simply clicking on a column heading. For example, in the directory listing, you can click on the City/State column to sort people in the list alphabetically by their stated city and state. Click on the column heading again to do a reverse sort.

At the very bottom of the NetMeeting window (yes, way down there) is the status bar. If you don't see it on your NetMeeting window, make sure that a check mark appears next to the Status Bar option in the View menu. The status bar displays messages telling you what's happening with NetMeeting, such as whether you're in a call or whether you're logged into an ILS server.

Clicking on different tabs changes not only the listing, but also the toolbar.

Getting Yourself Configured

Over the past year or so, the greatest strides in improving the user friendliness of Internet telephony programs have come through the wholesale adoption of Install and Configuration Wizards. These Wizards lead you, step by step, through the previously tricky set-up procedure for these programs by asking you simple, plain-language questions — most of them multiple guess (oops, choice). By simply answering these questions (such as "How fast is your modem?"), you get the configuration details — often the trickiest part of using Internet telephony programs — out of the way in a painless manner.

After you follow the Wizard and answer its questions, you're ready to make and receive calls as soon as you start NetMeeting for the first time. You can still mess around with plenty of Options and Preferences dialog boxes in NetMeeting, though, if you like to do that sort of thing. We could probably spend a whole book exploring the details of these dialog boxes. Instead, we focus on a few settings that you may actually need to access and change on occasion:

1. **Choose Tools⇨Options from the menu bar.**

   The Options dialog box opens, as shown in Figure 9-2. By clicking on the tabs at the top of the dialog box, you gain access to different categories of preferences that you can change.

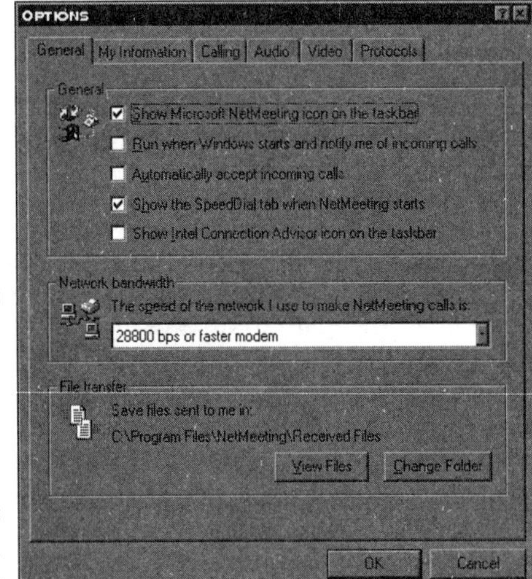

Figure 9-2: You can configure NetMeeting using the Options dialog box.

2. **If you are not already there, click on the General tab.**

   Most of these items are a matter of personal preference, but you need to watch out for a couple of them.

3. **In the General section of this tab, specify whether you want NetMeeting to answer all incoming calls without your intervention.**

   You activate the Automatically accept incoming calls option by clicking on its check box. (A check mark indicates that you have turned on this option.) We recommend that you don't turn on this option — unless you are a very friendly person, or if you log onto the server privately (we tell you how in a moment), in which case you only receive calls from people you know.

4. **Still on the General tab, verify that you selected the proper speed for your Internet connection when you ran the Setup Wizard.**

5. **Click on the My Information tab.**

   The Options dialog box changes to show the My Information preferences, as shown in Figure 9-3.

6. **Make sure your listing information is correct.**

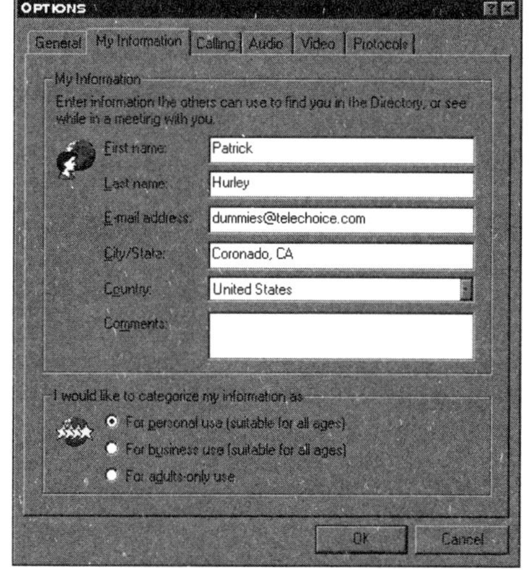

Figure 9-3: Enter information about yourself on the My Information tab.

Notice the three radio buttons at the bottom of the dialog box, under the section labeled I would like to categorize my information as. Click on the option that best suits your interests and the information that you've entered.

The For adults-only use button doesn't just mean that you won't have any kids calling you — if you get our drift. Just like Internet Chat, Internet telephony has its X-rated moments.

7. **Click on the Calling tab.**

   The Calling Options appear in the dialog box, as shown in Figure 9-4.

8. **In the Directory section of the Calling tab, make sure that a check mark appears in the check box labeled Log on to the directory server when NetMeeting starts.**

   If you have a dial-up Internet account, no one can find you and call you unless you select this option.

9. **Select the Server of your choice from the Server name drop-down list.**

   Or if you know the server's name, you can enter the name directly into the Server name text box. You may do this, for example, if you use NetMeeting in a corporate intranet environment and you have your own server. (You have to get the address from your network folks.) Additionally, a growing number of third-party servers are being launched, often catering to special interests — for example, perhaps you could log into the ILS that caters to Chargers fans, or one for parents of twins.

The ILS server that you log into is the one on which you are listed in the directory listing and Web phone books, for anyone who may be looking for you.

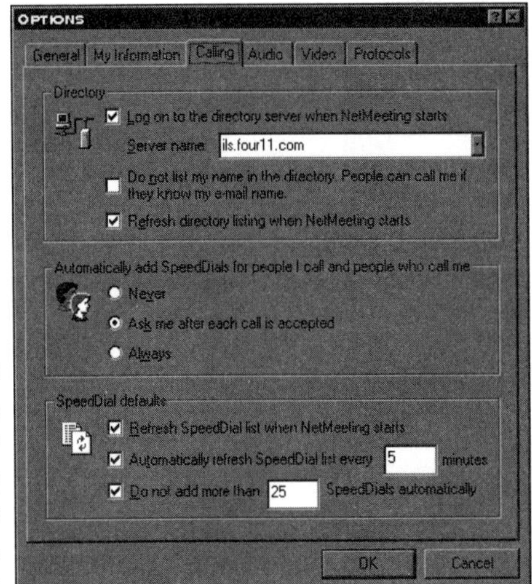

Figure 9-4: Select your calling preferences here.

10. **If you want to log onto the server privately (no listing in Directory), click on the Do not list my name in the directory check box.**

    As it says right there in the dialog box, you can still get calls in this mode — but only from people who know you and your e-mail address. Think of this option as an unlisted number of sorts.

11. **Below the Directory section, click on one of the three radio buttons that let you control automatic entries of people into the SpeedDial list.**

    We don't recommend the Always radio button, unless you want to have a huge SpeedDial listing. We tell you more about building your SpeedDial list in the section titled "Speed dialing," later in this chapter.

12. **Click on the Video tab.**

    Figure 9-5 shows the video options. (You can skip over the Audio tab — the Audio Tuning Wizard in the Set Up Wizard does a good job of getting you properly configured.)

13. **In the Sending and receiving video section, click on the check boxes to select or deselect the automatic sending and receiving of video when you start a call.**

    These options are really matters of personal preference — and no matter which settings you choose, you can start or pause video at any time during a call.

14. **In the Send image size section, click on the radio button corresponding to the size of the video window that you want to send to the other person in your call.**

    You can choose from Small, Medium, or Large — it's really up to you. Remember, however, that transmitting and displaying a larger video

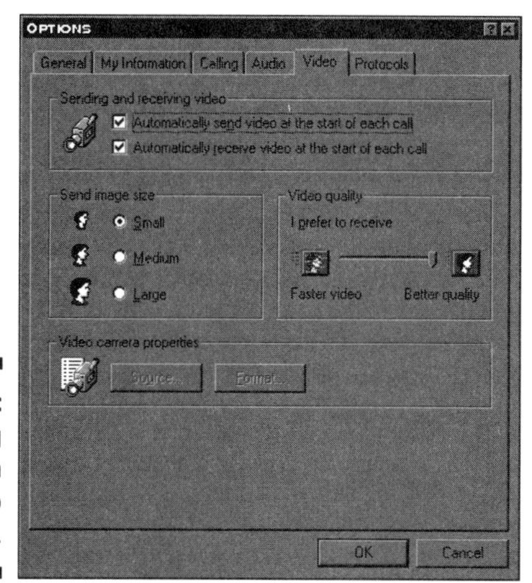

Figure 9-5: Messing around with your video options.

window requires more bandwidth, screen space, and CPU horsepower. We usually stick with the default: Medium.

15. **In the Video quality section, use your mouse to drag the slider to your preferred setting.**

    Faster video is more compressed, with an attendant loss in picture quality, but it can actually move on a modem-speed connection, whereas Better quality gives you a sharper picture, but with a reduced frame-per-second rate. It's your choice. Please note that this setting controls the video that you RECEIVE. The other party in the call controls how your video is received, not you.

16. **Click on the Protocols tab.**
17. **Make sure that a check mark appears in the Network (TCP/IP) check box.**

    Otherwise, you won't be able to place calls over the 'Net or over your corporate intranet.

Okay, wasn't that fun? Time to make a call and stop messing with preferences for a while.

How Do I Make a Call?

Placing a call in NetMeeting is a simple process. You can initiate calls over the Internet in one of several ways:

- From the ILS directory listing
- Directly, by entering an e-mail address or an IP address
- From your SpeedDial or History listings

You can also place calls from a Web directory. A whole bunch of these directories exist (one for each ILS server), and they contain the exact same listing of names as the directory listing inside NetMeeting. Calling from a directory varies a bit from directory to directory, but basically it involves clicking on a link on the Web directories page in your browser. (In addition to making calls via the 'Net, NetMeeting can make calls directly from modem to modem or over a corporate IPX network.)

Placing a call by using the ILS Directory

To initiate a call to someone by using the ILS directory listing, do the following:

1. **Click on the Directory tab in the main NetMeeting Window.**

   NetMeeting displays a listing of all the people in your ILS server. (To see an example of this listing, check out Figure 9-1, earlier in this chapter.)

2. **Scroll through the listing until you find the person you want to call.**

   You can sort the listing by clicking on one of the column headings in the listing. You can also narrow the number of entries in the listing by choosing a category from the Category drop-down list at the top of the directory listing. Which categories are available depends on how you categorized yourself in the My Information tab of the Options dialog box — for example if you log into the server's Personal category, you can sort through the listings of other Personal users. Generally speaking, you can view all entries in the directory, or subcategories — like Personal with video (which shows everyone in the Personal category who has a video camera), or Business Not in a call (which shows those users who categorized themselves as business users and are not currently engaged in a call). When you select all, you see everyone on the server, regardless of category, but in the other listings, you see only those people who have chosen the same category as you.

3. **If you want to view the listings on other ILS servers, just select a different one from the Server drop-down list.**

   The new list appears in your directory listing as soon as it is downloaded to your computer. (This can take a few seconds.)

4. **When you find the person you wish to call, just double-click on that person's entry in the listing.**

5. **NetMeeting communicates with the server to get your party's IP address, and then connects you.**

Calling directly

You can also call someone directly, by e-mail address or IP address, if you like. If you attempt to call someone by e-mail address, NetMeeting looks on the ILS to find the IP address for your party's computer and then connects the call. The steps for making a direct call are quite simple:

1. **Choose Call⇨New Call from the menu bar.**

   NetMeeting displays the New Call dialog box, shown in Figure 9-6.

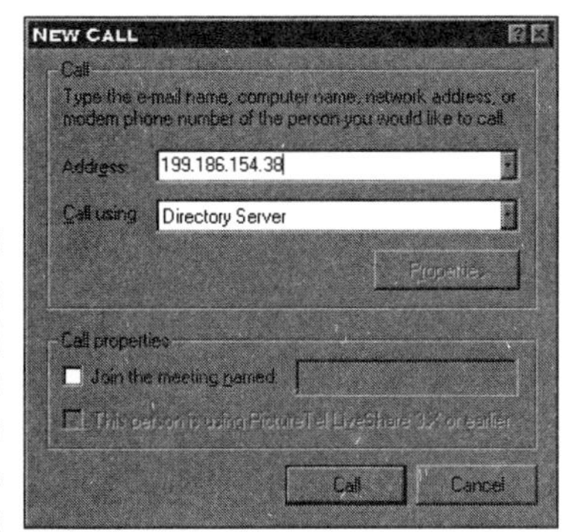

Figure 9-6: Enter call information in this dialog box.

2. **In the Address text box, enter the e-mail address, IP address, or computer host name of the person you wish to call.**
3. **From the Call using drop-down list, select automatic.**

   Note: You *can* specify TCP/IP or Directory server here, but why not let NetMeeting figure it out for you?
4. **Click on the Call button.**

NetMeeting determines what kind of call you are making, figures out an IP address (if it needs to), and rings the other person's NetMeeting.

Speed dialing

With the NetMeeting SpeedDial feature, you can easily find the people you usually call. No longer do you have to scroll through the Directory listing or guess at your friends' e-mail addresses. Just SpeedDial them! But before you start using SpeedDial, you need to figure out how to get people into your listing.

By default, NetMeeting automatically enters everyone you call into your SpeedDial list. As we mention in the section on configuring NetMeeting (earlier in this chapter), we're not too fond of that default — you can quickly get an unmanageable number of people in your SpeedDial list if you like to use NetMeeting to chat with strangers.

We prefer a different setting in the Options dialog box: Ask me after each call is accepted. With this setting, a dialog box pops up on your screen and lets you choose whether to add the person to your SpeedDial list. You can also add people to your SpeedDial list manually. Just go into the Directory, Current Calls, or History tab of the NetMeeting main window and

1. **Select a name in the listing by clicking on it.**
2. **Click on the SpeedDial button on the toolbar.**

   Or if you're on the Current Call tab, you can also choose SpeedDial⇨Add SpeedDial from the menu bar. NetMeeting opens the Add SpeedDial dialog box, shown in Figure 9-7.
3. **Click on the Add to SpeedDial list radio button.**
4. **Click on OK to save the SpeedDial listing.**

You can do some neat SpeedDial tricks with those radio buttons. For example, if you want to have someone call you, you can create a SpeedDial listing of yourself and then click on the Send to mail recipient radio button. This opens the Internet Explorer Mail program, and you can send the SpeedDial listing as an attached file. The recipient can retrieve this file and, by double-clicking on it, launch NetMeeting and call you. This trick is great for letting someone know that you'll be online and want to talk at a certain time. After you save some SpeedDial listings, you can place a call from the SpeedDial list:

1. **Click on the SpeedDial tab.**

   You see all of your saved SpeedDials. The Status column even tells you whether they're online and logged into an ILS server!
2. **Double-click on the name of the person you wish to call.**

They don't call it SpeedDial for nothing.

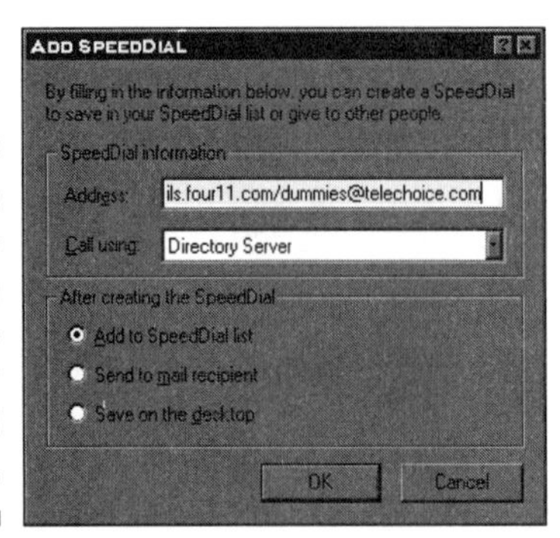

Figure 9-7: Manually adding someone to your SpeedDial list.

Incoming!

If someone calls you while you're running NetMeeting, you hear a ringing sound, and a dialog box like the one in Figure 9-8 pops up on your screen. Click on Accept to answer the call, or click on Ignore if you'd really rather not take a call.

Figure 9-8: Someone wants to speak with you.

You can turn off the ringer, so to speak, by choosing Call⇨Do Not Disturb from the menu bar. The first time you do this, NetMeeting displays a dialog box explaining to you the dire consequences of invoking Do Not Disturb. (You can't receive incoming calls.) If you don't mind seeing this reminder every time you choose this option, just click on OK. Or, you can turn off this reminder by clicking on the Don't show me this message again check box before you click on OK.

Making adjustments

When you're in a call, regardless of how it was initiated, NetMeeting automatically shifts over to the Current Call tab, as shown in Figure 9-9. From this strategic position, you can control all of the possible permutations of a call — from video to audio to text chat to heavy-duty collaboration.

Video games

When NetMeeting displays the Current Call tab, you probably focus first on the two video windows on the right-hand side. Depending on whether you or the other party have video and what video preferences you've both selected, you may or may not see anything in these windows right away.

If your video isn't set up to start automatically, just click on the triangular Start button on the My Video window to start sending. Clicking on the square Pause button will, of course, pause the video. If the other person is sending video but you're not configured to receive it automatically, just click on the Start button on the Remote Video window to begin receiving the other party's video.

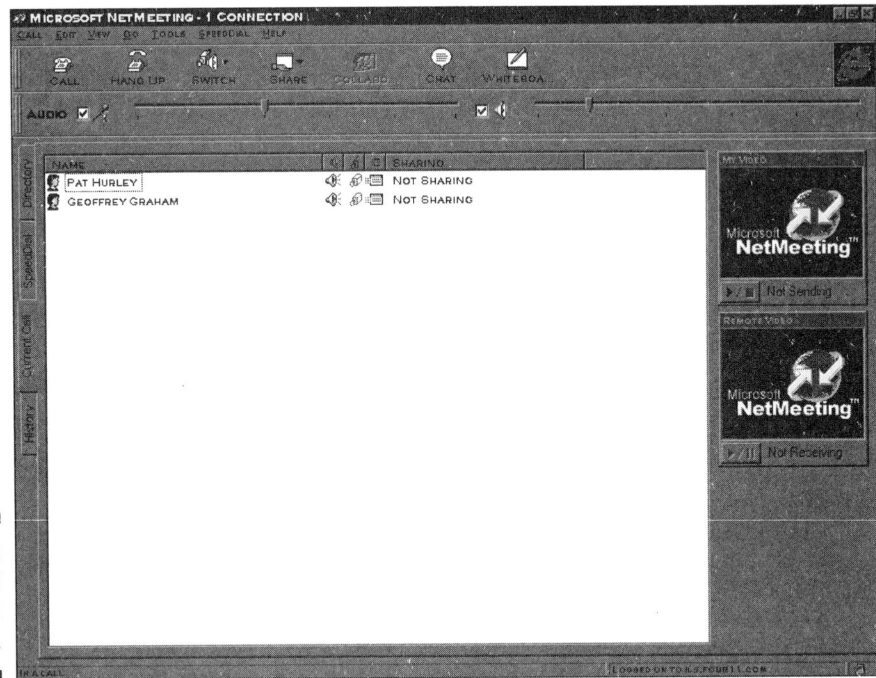

Figure 9-9: The call command center.

You can open the Video tab of the Options dialog box at any time by right-clicking on either video window and choosing Properties from the pop-up menu that NetMeeting displays. The options on the Video tab let you make adjustments to your video, like window size or incoming video quality.

You can also right-click on either video window and choose the Detach From NetMeeting command from the resulting pop-up menu. This command lets the video window break loose its bonds and become a free-floating, standard, Windows 95 window on your screen — one that you can move, resize, or even minimize down to the task bar if you'd like. Figure 9-10 shows a detached My Video window.

Figure 9-10: Lost in space — a detached video window.

Audio adjustments

Of course, Internet telephony involves more than just watching video. The bottom line is audio. Adjusting your audio during a NetMeeting call is pretty darn intuitive. Two check boxes — one each next to the microphone and speaker icons on the audio controls toolbar — activate the device when selected and mute it when not selected. Move the sliders left and right to decrease or increase the levels of your microphone or speakers.

When you talk, you should see a graphical representation on screen — some small, concentric semicircles radiating out from next to the microphone icon. You see something similar next to the speaker icon when the other party speaks.

If you have troubles with your audio, choose Tools⇨Audio Tuning Wizard from the menu bar and follow the Wizard's directions for reconfiguring your audio.

Breaking up is hard to do (Hanging up isn't)

Whenever you want to end a call, just click on the Hang Up button on the toolbar.

What Else Can I Do with NetMeeting?

You can do lots more in NetMeeting than just yak away. You can also have text chats, share images on a whiteboard, send files, and share applications. The first three items in this list have become common on most on Internet telephony programs, but application sharing is unique to NetMeeting. To start these functions, click on their corresponding buttons in the toolbar (while in the Current Call tab) or choose Whiteboard, Send File, or Chat from the Tools menu.

Application sharing

Application sharing is really the neatest trick that NetMeeting has up its sleeve. After you're connected to someone else's computer (or to several computers, if you're in a multiparty meeting — which we discuss in a moment), you can open any Windows application and have other people in the conference view it on their screens. The other participants don't even need the same application on their computers — so you can share, for example, a Microsoft Word document with a colleague who uses WordPerfect. Pretty neat.

Application sharing is not the same thing as whiteboarding. On a whiteboard, each conferee can view the same document but can make changes or mark up only an *image* of the document. In application sharing, people on the remote end of the call can actually take control of the program on the computer with the shared application and make changes in the document themselves! Assuming you are already connected in a conference and in the Current Call tab, here's how you share an application:

1. **Open the application you want to share.**
2. **Click on the Share button on the toolbar (or choose Tools⇨Share Application from the menu bar) and select the application from the list that appears.**

   The same program window you see on your screen appears on the remote party's screen — but they can't do anything with it, yet.
3. **If you want to let the other party use the application, choose Tools⇨Start Collaboration or click on the Collaborate button on the toolbar.**

   The other person, or people in a group meeting, must also do this step if they wish to collaborate. Only one party at a time can edit the shared document, so if the other party is editing, you need to wait until that person finishes before you can resume editing the document. Users can take control of a program by double-clicking on its window.
4. **If you want to go back to letting the other person see but not touch your program, click the Collaborate button again.**
5. **When you're finished sharing, click on the Share button on the toolbar again and select the application you want to stop sharing.**

Application sharing is a really powerful tool that gives you unprecedented capabilities for collaborating over the 'Net. Know this, however: Sharing an application requires a pretty big chunk of your computer's memory, bandwidth, and computing horsepower, so don't be surprised if this great feature slows things down a bit.

File transfer

NetMeeting also allows you to transfer any file from your computer to the other person in your call. To do this (while already in a call):

1. **Choose Tools⇨File Transfer⇨Send File.**

   A standard Windows File dialog box opens, as shown in Figure 9-11.
2. **Browse through your hard drive and select the file you wish to transfer.**
3. **Click on the Send button, and away the file goes.**

Figure 9-11: You can pick any file on your hard drive to transfer via the File Exchange.

Transferring a file takes some bandwidth (as anyone who has ever downloaded a 12MB Web browser file knows well), and because you are already using a lot of your available bandwidth for audio and video, things can really get bogged down when you add a file transfer to the mix.

You can access the files you receive by opening the Received Files folder on your hard drive. Choose Tools⇨File Transfer⇨Open Received Files Folder.

By default, NetMeeting puts the files in a folder within the NetMeeting Folder on your hard drive. You can select whatever destination you like in the Options dialog box, within the General tab.

Whiteboarding

The whiteboards included with Internet telephony programs like NetMeeting are a lot like the physical whiteboards you find in offices or schools. You can paste pictures or images onto the whiteboard and then use various drawing tools to mark up or annotate the base images. You can open the whiteboard in one of two ways:

- From the Tools menu, by choosing Whiteboard
- From the toolbar in the Current Call tab, by clicking on the Whiteboard button

The Whiteboard, shown in Figure 9-12, has many more features than we have space to talk about here (it could be a chapter of its own), but here's the bottom line:

- The menu bar at the top of the whiteboard lets you save and open whiteboard files, gives you access to the whiteboard tools, and lets you configure the whiteboard.
- Running down the left-hand side of the whiteboard, the toolbar contains standard drawing tools (like you'd find in Paint), plus a few whiteboard-specific tools for doing things like copying windows or screen areas to paste onto the whiteboard.

Figure 9-12: The whiteboard — dog picture is optional.

- The whiteboard itself — your blank canvas — takes up most of the whiteboard window.
- On the bottom right of the whiteboard, you find controls for adding and navigating the pages of the whiteboard — you're not limited to just one page of smiley faces and pictures of your dog!

We're not going to get into how to use the whiteboard — it's really simple, and fun, to figure out on your own. However, we do have two tips to offer for using the whiteboard:

- **Locking the contents:** The whiteboard, by default, is a *shared* whiteboard. In other words, anyone in a call or a meeting can mark it up, erase things, and so on. If you put something on the whiteboard that you don't want anyone to mess with, choose Tools➪Lock Contents. When you're ready to let them at it again, just repeat this process to deselect the lock feature.
- **Keeping things private:** Again, due to the shared nature of the whiteboard, everything you put on the whiteboard is seen by everyone else. If you have something you want to work on privately before letting anyone else see it, just choose Tools➪Synchronize from the whiteboard menu bar, to deselect this menu item. When you want to start sharing again, repeat this action, to reselect Synchronize.

Text chat

The fact that NetMeeting has built-in audio conferencing capabilities doesn't mean that you find no use for good, old-fashioned text chat. Text chatting is useful if one or both parties don't have sound cards or if you have trouble getting a good audio connection. Text chat is also essential in the multiparty conferences, or meetings, which we discuss in the next section of this chapter. (Because you can share audio and video with only one person at a time, text chat is your next best way to communicate with the rest of the people in the meeting.) Here's how to use text chat:

1. **While in a call, from within the Current Call tab, click on the Chat button on the toolbar (or choose Tools⇨Chat from the menu bar).**

   NetMeeting opens the Chat window, shown in Figure 9-13. The Chat window's title bar lists the number of people using Chat in your call or meeting. Using the Chat window is really quite intuitive.

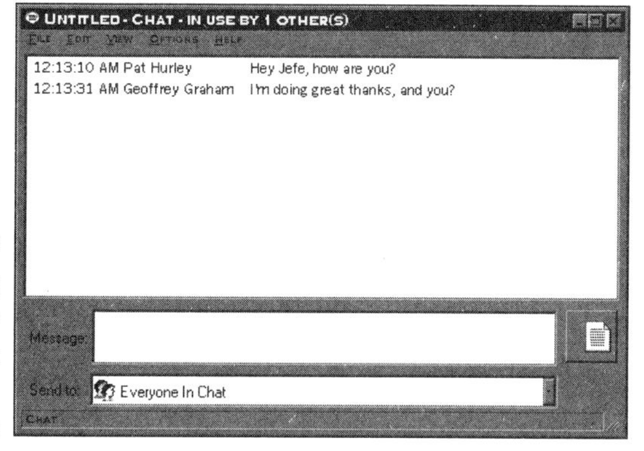

Figure 9-13: Type away in the Chat window.

2. **Type into the Message text field, and press the Enter key or click on the Enter button (which you find directly to the right of the Message text field — it looks like a piece of paper taking flight).**

   If several people attend your meeting, you can send text chat privately to any member of the meeting by simply selecting that member's name from the Send to drop-down list at the bottom of the window.

3. **Your text is sent and appears in the main text field of the Chat window, along with any incoming chat.**

After you've finished text chatting, close the Chat window by clicking on the Close button on the far right-hand side of the window's title bar or choose File⇨Exit from the menu bar in the Chat window. Before closing the window, NetMeeting displays a dialog box to prompt you to save the contents of the Chat. Click Yes to save, or No to discard the chat contents.

It's a Party: Multiparty Conferences

Previous sections in this chapter focus primarily on the person-to-person Internet telephony capabilities of NetMeeting. Microsoft designed NetMeeting with more than just that in mind, however. You can also use NetMeeting as a multiparty data conferencing tool. Getting a group of people together this way in NetMeeting is called — of course — a meeting.

What you can and can't do in a meeting

During a multiparty meeting, you can do everything we discuss in the preceding sections, including:

- Audio and video conferencing
- Application sharing
- Whiteboarding
- Text Chat
- File transfer

What you can't do, however, is share audio and video with everyone in the meeting. You can do this with only one person at a time. However, you can switch your audio and video connection to different people in the meeting, one at a time. We tell you how in just a second.

Hierarchies, and stuff like that

It's worth taking a few seconds here to describe how NetMeeting organizes meetings — the hierarchy, if you will, of a meeting:

- **The Conference Host:** This is the person who initiates the first call in the meeting. Think of the Conference Host (or top provider, as Microsoft describes this person in its documentation) as the cornerstone of the conference. Without the host, the conference would fall apart.
- **The participants:** This category includes everyone else in the meeting. Participants can join the meeting by calling (or being called by) the conference host, or they can join by calling (or being called by) another participant who has already joined the meeting.

Because of the way data flows in NetMeeting conferences, the person that you connect with to join a meeting is important. Here's an example: Suppose that five people are conferencing. Pat is the conference host, and Rebecca and the other participants join the meeting when Pat calls them. You (should we call you Danny?) can join the conference by calling (or being called by) either Pat or Rebecca (or any the other participants, for that matter). If you join the meeting by calling Rebecca, all your data to the rest of the meeting participants flows through Rebecca's computer and then through Pat's to the rest of the group. In other words, data doesn't flow directly from your computer to every other computer in the meeting — instead, it moves from one computer to the next in the connection.

Why do you care how your data flows? Well, you really don't — most of the time. But if you connect to the meeting through Rebecca instead of Pat, and Rebecca decides to leave the meeting, your connection goes away with her. If you had been connected to Pat instead, you would remain in the meeting.

Your best bet when joining a meeting is to make your connection directly with the conference host.

How to set up a meeting

You don't really have to do anything special to set up or join a meeting. If you try to call someone who is currently in a call, NetMeeting pops up a dialog box telling you that the other person is in a call and asks you if you want to join the meeting. Click on Yes, and if the other person accepts you, you're in a meeting! Likewise, if you are in a call and someone else calls you (and asks to join your meeting), you see a dialog box saying that this person wishes to join your meeting. Click on Accept to let the person join, and you've got an instant meeting.

If you call someone in a meeting who hangs up, NetMeeting disconnects you not only from the person you called, but also from the rest of the meeting.

You can set up meetings in a more formal way, if you'd like. Give advance notice of the meeting time and host to everyone whom you'd like to have join the meeting. If you plan to be the host, here's how you get things started:

1. **Connect to the Internet.**
2. **Start NetMeeting.**
3. **Choose Call⇨Host Meeting from the menu bar.**

The other participants, for their part, simply place calls to you in any of the normal fashions (Directory, SpeedDial, or Direct). After everyone connects to you, the meeting has begun.

You don't really need to do anything during a meeting. Most NetMeeting functions — such as text chat, whiteboarding, file exchange, and application sharing — work exactly as they do in a person-to-person call. The only difference you may notice (besides the extra folks talking in the Chat window, or the extra smiley faces being drawn on the whiteboard) is that things slow down a bit — the data takes more time to get around to everyone in the meeting, and you probably have more data moving around than you do in a regular call.

The major exception to this "all conditions normal" situation involves the sharing of audio and video. Remember that these functions are strictly limited to a person-to-person mode only. However, that limitation doesn't mean that only two people in a meeting can use audio and video at once — just that you can share audio and video with only one person at a time. If a meeting has six participants, all of them may be using audio and video at once, but in three separate and unconnected conversations.

At any time during a meeting, you can switch the person with whom you share audio and video. NetMeeting lets you make the switch in any of the following ways:

- From the Tools menu, choose Switch Audio and Video⇨the name of the person to whom you want to switch.
- In the toolbar, click on the Switch button and then select the name of the person with whom you want to share audio and video.
- In the Current Call tab, right-click on the name of the person with whom you want to communicate and then choose Switch Audio and Video from the pop-up menu that NetMeeting displays.

Putting NetMeeting to Bed

Life can't be all fun and games. At some point, you have to shut down NetMeeting and leave work. We know that this is hard for you to do (the leaving work part), but at least it's easy to shut down NetMeeting. Just choose Call⇨Exit from the menu bar. Checking out with the boss is up to you — next time, use NetMeeting as a telecommuter.

Chapter 10

Using the VocalTec Internet Phone

In This Chapter

- Setting up Internet Phone for your system
- Logging onto the server
- Picking a Chat Room
- Making and receiving calls
- Getting multimedia savvy

VocalTec's Internet Phone is not (yet) an H.323-compatible phone, but we want to talk about it anyway because this program is probably the most popular Internet telephony program as of this writing. If you can't find someone to talk with by using Internet Phone, you're probably not trying.

Internet Phone is also fairly easy to use and has an excellent Help system, so you shouldn't have any problems getting the program up and running. Our goal is to keep you from ever needing to hit that Help button, but having it is nice — just in case.

The version we discuss in this chapter, Release 4.5 for Windows 95, is the most advanced version of Internet Phone — it includes a videoconferencing capability. A similar version, without video, is currently available for Windows 3.1, and a version (also without video, as well as several other features) is also available for the Macintosh. Release 4 for Windows 95, Version 4.0 for Windows 3.1, and the Mac version are all on the CD-ROM that accompanies this book.

A Quick Peek at the Interface

Figure 10-1 shows the main Internet Phone window. Before you start doing anything, take a look at where everything is located.

Figure 10-1: The main Internet Phone window.

You find the following six pull-down menus across the top of the window:

- **Phone:** Contains the general controls for placing and hanging up calls and for opening the Voice Mail, Whiteboard, and Text Chat features.
- **View:** Contains controls for setting up system preference options and for customizing your view of information about calls and people.
- **Call Center:** Provides access to the different methods Internet Phone offers for finding and calling people.
- **Audio:** Contains controls for Internet Phone's audio functions.
- **Video:** Lets you control the sending and receiving of video.
- **Help:** Provides access to the Help function.

On the left side, below the menu bar, Internet Phone offers the following four buttons, called *Session Tools* (from top to bottom):

- **Voice Mail:** Opens the Voice Mail dialog box.
- **Text Chat:** Opens the Text Chat window.
- **Whiteboard:** Opens the Whiteboard window.

  Note: A *whiteboard* is a shared window on which you can draw or paste graphics and text to exchange with another person.

- **Transfer File:** Enables you to send a file to the person with whom you're talking.

Next to the Session Tools is the *Animated Assistant,* a little cartoon character who acts out what you are doing on Internet Phone. (We think he looks as though he's wearing Charlie Brown's shirt.) For example, if you're talking, you see him with a telephone in his hand, yakking away.

Directly beneath him is the *Status Line*. (In case you're wondering, we feel we should always refer to the Animated Assistant as *him,* not *it,* because he's clearly a male and he's just so friendly looking — too bad he doesn't have a name.) The Status Line tells you in words what the Assistant is doing — so if the Assistant is talking (which means, of course, that you are talking), the Status Line displays `Talking`.

To the right of the Assistant, you find the following three buttons, called the *Conversation buttons* (from top to bottom):

- **Answer:** Click on this button to accept an incoming call. If you're already on a call with someone else and a new call comes in, this button becomes the Call Waiting button.
- **Hold:** This button enables you to place your current call on hold (so you can make another call, perhaps, or just run to the fridge). After you put someone on hold, this button changes to become the Resume button — click on the button again to (you guessed it) resume your conversation.
- **Hang Up:** Click on this button to conclude a call.

Directly beneath these buttons are two more buttons, to start and stop the video. The one on the left controls your outgoing video, and the one on the right starts and stops the reception of incoming video.

Beneath the Assistant and the Status Line, you find the *Audio Setup Tools*. You don't actually use these tools to set up your audio configuration (we tell you how to do that in the section "Configuring Internet Phone," just a little later in this chapter) but rather to adjust and monitor your audio during a conversation. Internet Phone has the following Audio Setup tools:

- **Microphone Mute:** Click on this button to toggle your microphone on and off.
- **Record Level:** This bar graphically depicts your outgoing audio level by means of a simulated LED display.
- **Playback Level:** This bar graphically depicts your incoming audio level.
- **Playback Slider:** This tool consists of a little slider that you move back and forth with your mouse to adjust the playback volume. (The slider itself is the little upward-facing triangle located just under the Playback Level bar.)

- **Speaker Mute:** By clicking on this button, you toggle your speaker on and off.

Running across the screen, in the bottom part of the main Internet Phone window, are three buttons that show or hide the following Internet Phone panels:

- **Call Center:** This panel contains a window that displays the name of the user you're calling or with whom you're talking, plus three buttons that each lead you to one of the program's directory services. (We describe these services soon — but if you can't wait, they're the *Global Online Directory,* the *Personal Directory,* and the *Web Directory*). You can also enter an IP address or Internet Phone address (a specific address assigned to a registered user) to make a call directly to a person without going through a server.
- **Session List:** This panel displays the names of the people you called or who have called you during your current Internet Phone session. You can place calls to people directly from this list by double-clicking on their names.
- **Statistics:** This panel displays a moving graph of incoming and outgoing data packets and has numerical displays of the percentage of packets lost and any delay times. The panel is pretty neat to watch and can help you see what's going on with the quality of your call.

In Figure 10-1, we have all three of these panels open. You can close the panels, if you like, by clicking on the buttons. If the triangle at the right of each button faces downward, the panel is hidden; if the triangle faces upward, the panel is open.

Besides the main Internet Phone window, you can also access a second window, called the *Internet Phone Global OnLine Directory.* (This name is certainly a mouthful — and a lot to type — so we just call it the OnLine Directory from here on to avoid any repetitive stress injuries.) The OnLine Directory (shown in Figure 10-2) opens automatically after you start Internet Phone and appears on the right of your screen, next to the main window.

Across the top of this window is a menu bar containing the following four menus (from left to right):

- **Directory:** Enables you to manually connect and disconnect from the server and to refresh the listing of Chat Rooms.

  Note: *Chat Rooms* (called *Topics* in previous versions of Internet Phone) are virtual "rooms" (arranged alphabetically by subject) that you can "enter" to find people to talk with. After you enter one of these rooms, your name appears on a list of people "in" the room — so you can call and be called by others on the list.

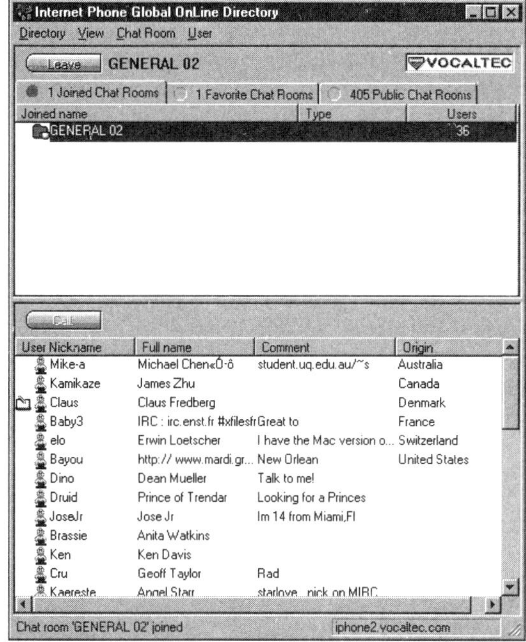

Figure 10-2: Use the OnLine Directory to find folks to talk to on Internet Phone.

- **View:** Enables you to customize your view of the Directory and its Chat Room listing.
- **Chat Room:** Enables you to enter and leave Chat Rooms.
- **User:** Enables you to call someone in a Chat Room, get more information about the person, or add that person to your Personal Directory (which we describe in the section "Using your Personal Directory," later in this chapter).

Beneath this menu bar is a Leave/Join button that enables you to leave or join a particular Chat Room. (You read more about Chat Rooms later in this chapter.) Which version of the button you see depends, of course, on whether you're already in the Chat Room. Directly beneath this button are the following three tabs (from left to right) that toggle the contents of the listings of Chat Rooms and of the people in the Chat Rooms:

- **Joined Chat Rooms:** Shows the listing of the Chat Room you have joined.
- **Favorite Chat Rooms:** Shows the list of Chat Rooms that you automatically join when you log onto the server.
- **Public Chat Rooms:** Shows the listing of all public Chat Rooms on the server.

Clicking on one of these three tabs moves that tab to the foreground (they appear in 3-D) and lights up the red button on the left-hand side of the tab.

Below the tabs is a listing of Chat Rooms — which rooms appear in the list depends on which tab you select. Clicking on the name of a room in this list places that Chat Room name next to the Leave/Join button.

The bottom of the OnLine Directory window contains a listing of everybody in whatever Chat Room you select from the listing of Chat Rooms. You can sort this list of people by clicking on one of the buttons at the top of the list. (You can sort by User Nickname, Full Name, Comment, or Origin — where they're calling from.) A Call button also appears in this section of the window, enabling you to place a call to any person you select from the list — just click on the name and then click on the Call button.

You can change the relative size of the Chat Room list and the User List in the OnLine Directory by clicking and dragging the line that separates these areas of the window.

Configuring Internet Phone

When you first install Internet Phone, you encounter a Setup Wizard. (Wizards are programs that lead you through a complicated procedure — we love them.) This Wizard prompts you to provide some of the information you need for configuring Internet Phone. Specifically, the Wizard asks you to enter your personal information (such as your name and nickname), your SMTP server, and your ISP connection speed. (*SMTP,* by the way, stands for Simple Mail Transfer Protocol, and your SMTP server is provided by your ISP — check your e-mail program configuration or call your ISP's tech support to get this information.)

After you install the program and complete the Setup Wizard, you can start Internet Phone (by clicking on the Start button and choosing Programs⇨Internet Phone 4⇨Internet Phone Release 4), and you can finish setting up. Internet Phone automatically tests your computer to make sure that your machine is fast enough to run the program, and it runs an audio test to ensure that your audio configuration is correct. You can run the audio test again at any time (which you should do if you change sound cards or otherwise change your system).

Fine-tuning your audio

To run the audio test manually, follow these steps:

1. **Choose Audio⇨Audio Test from the menu bar.**

   The Audio Test dialog box opens, as shown in Figure 10-3.

2. **Click on the Start Test button and speak into your microphone.**

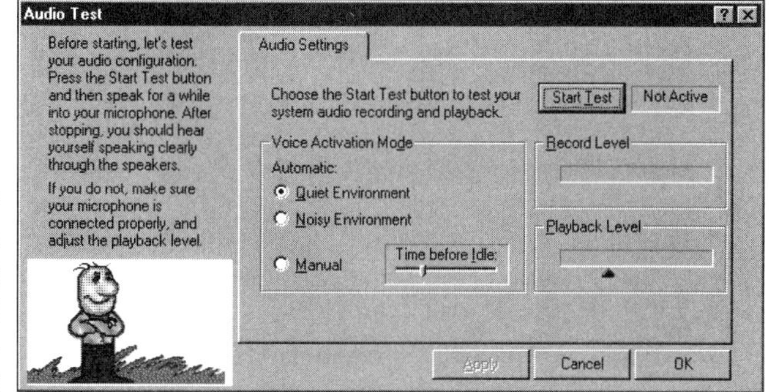

Figure 10-3: Testing your audio setup.

After a slight delay, you hear yourself talking back.

3. **Click on the Stop Test button (the Start Test button has changed!) to finish your test.**
4. **Click on the button next to Quiet Environment or the one next to Noisy Environment (depending on how noisy the room your computer lives in is) to set the Voice Activation Mode to Automatic.**

   These buttons choose a minimum level of sound to activate your microphone and start sending your voice across the Internet. You can also click on the Manual button if you want to manually turn your microphone on and off while you speak — although we never do this, because doing so is sort of a pain and the automatic activation works very well.
5. **Leave the Time before Idle slider in its current position.**

   Later on, you can go back and adjust this slider if you want. If you find yourself getting cut off when you pause between words, you can increase the idle time (or you can decrease the time if you stop talking but your microphone stays on too long).
6. **Click on the Apply button to save your configuration.**
7. **Click on OK to finish.**

Setting up your video

Even though the Setup Wizard takes care of most aspects of your configuration, you may want to check a few things before you start making calls with Internet Phone if you have a video camera. For example, you need to decide whether you want to videoconference automatically when you connect on a call or whether you want to start sending or receiving video manually. You can easily take care of these set-up options:

1. **Choose View⇨Options from the menu bar.**

   Internet Phone opens the Options dialog box, shown in Figure 10-4.

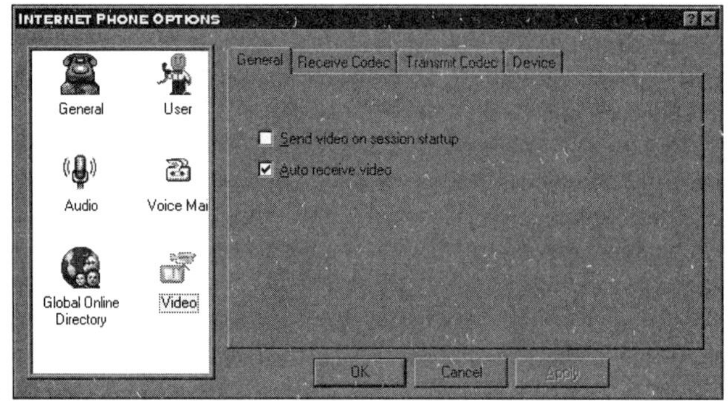

Figure 10-4: Setting your video options.

2. **Click on the Video icon and then click on the General tab.**
3. **Click on the Send video on session startup check box if you want to automatically send your video upon connecting a call.**
4. **If you want to receive video automatically when you connect, click on the Auto receive video check box.**

   Beware because you never know what may pop up in some complete stranger's video when you connect. Forewarned is forearmed!

   While you're in the Video Options dialog box, you may want to mess with a few more settings.
5. **Click on the Transmit Codec tab.**
6. **You should use the default codec (VocalTec Video Codec (vvc 1)), so don't mess with this setting.**

   You may have other codecs installed in your system that you can mess around with later on, when you feel comfortable with fine-tuning your system's video capabilities, but you really should use the vvc codec with Internet Phone.
7. **Click on the Configuration button.**

   The VocalTec Video Codec Configuration dialog box opens.
8. **Click on the Compress to color radio button if you have a color camera and video digitizer; otherwise, leave this option at the default setting, Compress to gray.**
9. **Click on the OK button.**

   The VocalTec Video Codec Configuration dialog box closes, and you're back in the Transmit Codec tab of the Video Options dialog box.
10. **Click on the Test button.**

    The Codec Test dialog box opens on your screen, as does your Self View video window.

11. **Use your mouse to drag the Codec Quality slider to your desired setting.**

    Dragging the slider to the left (toward the 0% end of the scale) decreases the amount of bandwidth used by your video but also decreases the clarity of your picture. Moving the slider in the opposite direction has the (of course!) opposite effect. Modem users should probably keep the slider down on the 0% side of the scale — trading off some picture quality for a picture that actually moves on the other end of your call. Experiment on a few calls to find the setting that works best for you.

12. **Click on OK to close the Codec Test dialog box.**
13. **Click on OK one more time to close the Options dialog box.**

After completing this process, you're ready to send and receive video. The Video menu includes some commands that let you access the standard controls of your video digitizer. These commands (Image Size and Quality, Camera Adjustments, and Self View Adjustments) invoke the controls that come with your video card or parallel port camera — and depending on the equipment you have installed, you may not even have all these options available. We can't tell you exactly what to do here because the controls vary widely depending on what you have installed. This is one of those cases in which you just have to read the manual. Sorry!

You can also change the other configuration options of Internet Phone at any time:

1. **Choose View⇨Options from the menu bar.**

   Internet Phone opens the Options dialog box.

2. **Click on any of the icons on the left-hand side of the dialog box and change your settings for that category of options.**
3. **After you finish, click on the OK button to close the Options dialog box.**

In practice, you rarely — if ever — need to go into this dialog box, but if you want to change your nickname or if you get a faster ISP connection (if, for example, you get an ISDN line), you can make changes in the Options dialog box to make sure that you keep your Internet Phone up-to-date.

Making the Server Serve You

As Chapter 6 discusses, Internet Phone uses network servers, based on the Internet Relay Chat (IRC), to connect parties who want to speak with one another. (The actual connection goes point-to-point over the 'Net, not through the server.) This newest version of the program also enables direct calls to a person's IP address or Internet Phone Address, which we cover in a moment (in the section "Making a Direct Call," later in this chapter).

Connecting to the server

Your first step, therefore, in talking on Internet Phone is to log onto a server. Luckily, Internet Phone does this for you automatically. Here's how you handle your part:

1. **Establish an Internet connection (if you use a dial-up ISP connection).**
2. **Click on the Start button and then choose Programs⇨Internet Phone 4⇨Internet Phone Release 4.**
3. **Internet Phone automatically searches through its server network and connects you to a server.**

   The name of the server you're connected to appears in the bottom right-hand side of the OnLine Directory window. If you don't see it listed, choose View⇨Status Bar from the menu bar.
4. **You are automatically joined to one of the General Chat Rooms.**

You're connected and ready to start talking.

Internet Phone works this way when fresh from the showroom floor. If you have been playing around with the Options dialog box, however, you may have turned off the auto connect feature. To turn it back on, choose View⇨Options from the menu bar, click on the General tab, and make sure that the check box titled Open Global OnLine Directory on Startup is selected. If you turned off the auto connect feature on purpose, you can always join the OnLine Directory by simply clicking on the OnLine Directory button in the Call Center of the main Internet Phone window.

Getting into Chat Rooms

After you're connected to a server, you need to join one or more Chat Rooms to find someone to talk with. As we mention in the preceding section, you automatically enter one of the General Chat Rooms as soon as you connect to a server, but you're no doubt going to want to join even more. Internet Phone servers have two different kinds of Chat Rooms, *public* and *private,* which function as follows:

- **Public Chat Rooms:** These chat rooms are openly available to anyone on the server, and the names of everyone connected to the server appear on the Chat Room listing for all to see.
- **Private Chat Rooms:** These chat rooms don't show up on the listing of Public Chat Rooms and can be joined only by those with whom you share the name of the chat room (unless someone guesses the name randomly).

You can be joined to 10 Chat Rooms at once — in any combination of private or public. After you have 10 Joined Chat Rooms, Internet Phone does not enable you to join another without first leaving one. Joining Chat Rooms is not a permanent thing — don't be overly concerned about which ones you join, because you can always leave them and try others.

Switching between Chat Rooms in your Joined Chat Rooms listing is a simple procedure — just click on the name of the Chat Room you want to view. You're still actively joined to the others (and can receive calls from people in them), but you see only the listing of users for the room you selected.

Public Chat Rooms: talking up a storm

Joining a Public Chat Room is a simple four-step process:

1. **Click on the Public Chat Rooms tab in the OnLine Directory.**

   A listing of all Public Chat Rooms appears, as shown in Figure 10-5.

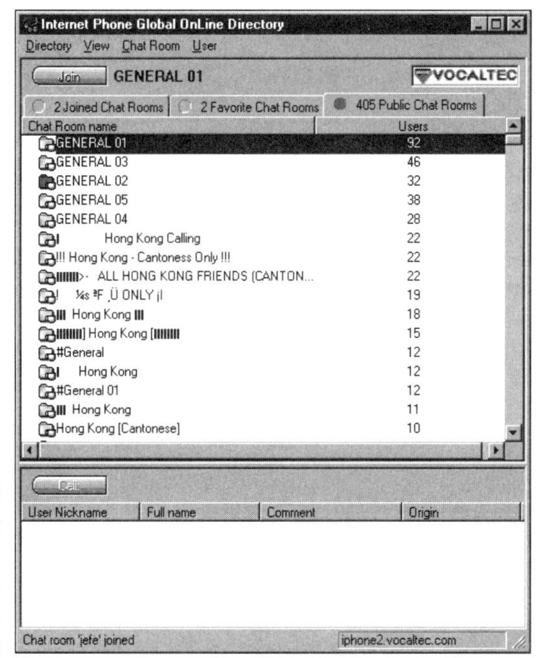

Figure 10-5: The Public Chat Room listing.

2. **Scroll through the list until you find a room you want to join.**
3. **Click on the name of the room.**

   The Leave/Join button changes to Join, and the name of the Chat Room appears next to the button.

4. **Click on Join, and you are in!**

   Click on the Joined Chat Rooms tab, and you see that you made it. The room you joined appears in the listing of Joined Chat Rooms below the tab.

If you want to create your own public Chat Room, you can! Just follow the steps in the section "Private Chat Rooms: Limited to friends, family, and associates," just a bit later in this chapter, taking care to perform the one different step that we note.

You find many different public Chat Rooms from which to choose. VocalTec created five permanent Chat Rooms; you see them listed as `GENERAL 01`, `GENERAL 02`, and so on. The public Chat Room areas created by members range from fly fishing to adults-only chat — and they change constantly.

You can join the General Chat Rooms set up by VocalTec only one at a time. If you try to join a second one, you're bumped out of the first. This limitation, however, is true only for the permanent VocalTec General Chat Rooms; you can join any of the other Chat Rooms you want, up to a total of 10.

Note: The permanent General Chat Rooms are an exception to the rule — even if you are already in 10 rooms (remember, that's your maximum number — public and private), attempting to join a second permanent General Chat Room bumps you out of the first and puts you in the second. You don't get a warning here that you're already in 10 rooms.

To leave a Chat Room, follow these steps:

1. **Click on the Joined Chat Rooms tab.**
2. **In the Chat Room list, click on the name of the Chat Room you want to leave.**

   The Leave/Join button changes to Leave.
3. **Click on the Leave button.**

   You're outta there.

Private Chat Rooms: limited to friends, family, and associates

Unlike the public Chat Rooms, which are visible to all users on the server, private Chat Rooms cannot be seen by anyone and can be joined only by people with whom you share the Chat Room name.

For example, you and a business associate can create a private Chat Room and always find each other quickly and easily by entering the Chat Room. (After you enter a private Chat Room, that room is automatically entered on your Favorite Chat Rooms listing, so you join the room every time you start Internet Phone.) No one else can join in without knowing the name, so you can be assured of some privacy; and if you join no public Chat Rooms, you receive no unwanted calls from strangers.

Whoa, Junior!

It's no secret: The Internet is still an anarchic, unruly, and, occasionally, adults-only place. Despite recent attempts by a few people — who probably don't understand it — the 'Net remains a mostly unregulated zone. Just as you can via the book store, cable television, and the old phone system, you can find sexually explicit material online.

And as is true of those other places, you find that, despite all the bad press, the vast majority of material on the 'Net does *not* fall into the pornographic category. If you've never run into adult material on the Internet, however, you very well may have your first experience on Internet Phone. A fair number of Chat Rooms on the servers have titles that we can't reprint here and that just may make you blush. Remember: You don't have to join them if you don't want to — skipping right over them is easy enough.

VocalTec is no doubt aware of this popular use of Internet Phone, and the company has created a way for you to hide these potentially offensive rooms from your view. Just follow this simple procedure:

1. **Click on the Public Chat Rooms tab in the OnLine Directory.**
2. **Click on the name of the room you want to hide.**
3. **Choose Chat Room⇨Hide from the menu bar.**

   Poof, it's gone.

Should you ever want to unhide this room, choose View⇨Hidden Chat Rooms from the menu bar. A dialog box opens, containing the names of all the rooms you've hidden. Click on the name of the room you want to unhide (we just made up that word, by the way, but it sounds good to us!) and then click on the Remove button. Click on the Close button, and the formerly hidden room is back on your list. (You need to wait until you reconnect to actually see the name on your list again, unless you choose Directory⇨Update Public Chat Room List from the menu bar.)

If you want to keep your private Chat Room really private, treat the Chat Room name as you'd treat a password. Don't select a Chat Room name that's too obvious and don't go around telling everyone the Chat Room name. If you choose *sex* as the name of your private Chat Room, you can probably be assured that someone else can figure out the name, and you soon find that your Chat Room is no longer private.

Setting up a private Chat Room is easy. Just follow these steps:

1. **In the OnLine Directory, choose Chat Room⇨New/Private from the menu bar.**

   Internet Phone displays the New Chat Room dialog box, similar to the one shown in Figure 10-6.

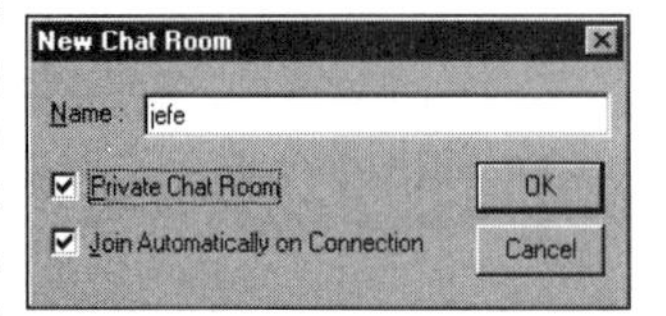

Figure 10-6: Creating a new Chat Room.

2. **Type the name of the Chat Room in the Name text box.**
3. **Click on the check box labeled Private Chat Room to select this option.**

   You can also create a new public Chat Room by following these very same steps but skipping Step 3.
4. **If you want to enter this private Chat Room every time you log onto the Internet Phone server, leave the Join Automatically on Connection check box selected.**

   This step places your private room onto your list of Favorite Chat Rooms. If you don't want to place the room on this list, click on the check box to deselect it.
5. **Click on OK to close the dialog box and enter the Chat Room.**

To join a private Chat Room set up by someone else, simply type the name of the existing Chat Room in Step 2 of this procedure and then continue on with the rest of the steps.

Initiating the Call

After you select a few Chat Rooms in which you're interested, you're ready to make a call. All you need to do is pick a person from within one of your Chat Rooms and click on that person's name. Here's how you do it:

1. **In the OnLine Directory, click on the Joined Chat Rooms tab to show the rooms you are in.**

   The names of all joined Chat Rooms (public and private) appear beneath the tab in the Joined Chat Room listing.
2. **Click on the name of the Chat Room that interests you.**

   The names of everybody in the Chat Room appear in the bottom part of the OnLine Directory's window, in the User List section.
3. **Click on a person's name and then click on the Call button (or just double-click on the person's name).**

You can also right-click on a person's name in the User Listing, which opens a pop-up menu that enables you to find out more information about the person, add the person to your phone book, or call that person.

Internet Phone determines the person's IP address and connects you.

Messing around with video

If you — or the person with whom you are talking, or both of you — have video, you can do a few things with video windows during a call. As far as your own video, you should make sure that your Self View video window is open — so you can monitor the video you send out. (You wouldn't want to end up with the camera aiming straight up your nose without knowing it, now would you?) Just choose Video⇨Self View from the menu bar, and poof, you can see your video! Don't forget to smile at the camera.

If you have a slower computer, especially if you are using a parallel port camera (which uses up a fair chunk of your CPU's capabilities to do its business), you may want to close your Self View window. Drawing your video picture on the screen is just one more task for the CPU — a task it may not want to do.

The Options dialog box enables you to set up Internet Phone to send video automatically when you make a connection, or to do it manually. Either way, you can change the status of your video transmission (from sending to not sending, or vice versa) by clicking on the left-hand video button on the main Internet Phone window or by choosing Video⇨Send from the menu bar.

If the other person in the call has video and is sending, you should see that person's video in place of the animated assistant. If the other person has set up a video for a larger size than can fit in the animated assistant's space on the Internet Phone window, the animated assistant disappears, and a separate video window opens on your screen. Like your outgoing video, you can toggle incoming video on and off by clicking on the right-hand video button on the main Internet Phone window or by choosing Video⇨Receive from the menu bar. You can also magnify the incoming video by choosing Video⇨Magnify Remote View.

Putting someone on hold

Internet Phone enables you to have multiple calls connected at one time — although you can talk with only one person at a time. You can put a call on hold by clicking on the Hold button at the top of the main Internet Phone window or by choosing Phone⇨Hold from the menu bar. After you place a call on hold, the name of the person you were speaking with moves down to the Session List section of the main window, and you can initiate or receive another call normally. If you want to return to your original call, simply click on the first person's name in the Session List.

Hanging up

After you finish the conversation, click on the Hang-up button on the Internet Phone main window or choose Phone⇨Hang Up from the menu bar to disconnect the call.

Finding out who's online

Trying to find someone in particular? Sure, you can join several of the Chat Rooms on a particular server and then scroll through the lists of people online to find the person, but if hundreds of people are online at once, how much fun is that?

If you're looking for a particular person, a directory assistance service makes all the difference in the world.

Web directory assistance

VocalTec has an online user directory of Internet Phone users on its Web site. Users who have either Netscape or the Microsoft Internet Explorer Web browser can go to the site directly by choosing Call Center⇨Web Directory from the menu bar in the Internet Phone main window. This command launches your browser and takes you directly to the user directory page. The directory has a complete alphabetical (by Internet Phone nickname) list of all users currently logged on. To place a call to someone from the Web Directory, simply click on the link containing the person's Internet Phone nickname.

You can also search for someone by Name, nickname, Host name (in other words, an Internet address — perhaps you'd enter **telechoice.com** if you were trying to find one of us online), or country.

Additionally, the Web Directory lets you look and see who's in certain chat rooms. This option works only for the five General Chat Rooms set up by VocalTec.

The addressing service

Registered users of Internet Phone (the Windows version anyway) have another option for making themselves accessible on Internet Phone. VocalTec runs an addressing service that basically performs the same function as the directory services utilized by many of the H.323 phones we discuss in this book. Part of the Internet Phone registration process (an optional part — so you don't have to register if you don't want to) is the assigment of an Internet Phone Address. After you've been assigned an Internet Phone Address (which looks just like an e-mail address), you can give it out to friends and associates whom you want to call you.

You can find directions for calling an Internet Phone Address a bit later in this chapter, in the section "Making a Direct Call." It's really easy to do, and because Internet Phone Addresses never change (unlike IP addresses), it's one of the best ways to get a hold of someone.

You can find someone's Internet Phone Address (if the person has one — remember that not all users are registered) by viewing that person's User Information dialog box. Just right-click the person's name in the Session List and choose User Info from the menu that pops up. The Internet Phone Address is the last item in the resulting User Information dialog box.

Here's another tip (we try to be helpful when we can): If you are a registered user of Internet Phone but declined to join the addressing service when you registered, you can change your mind at any time and join. Just choose Help⇨Registration Wizard from the Internet Phone menu bar. The Wizard starts up, and you just follow its directions. Couldn't be easier.

Saving Names and Making Calls

Looking through Chat Rooms and directories can be interesting and useful if you're trying to find someone, but what about those people with whom you've already talked and you are trying to find again? Internet Phone has a few ways of tracking these people down that may save you some time and effort. If you find yourself in this situation, try using one of the following features:

- **Personal Directory:** Your own phone book
- **Incoming and Outgoing History:** A listing of everyone you've called or who has called you
- **Session List:** A listing of everyone you've talked to during your current Internet Phone session

Using your Personal Directory

Internet Phone has a Personal Directory that acts as a phone book for you. You can save a name from the OnLine Directory, the Session List, or one of the history folders into your Personal Directory and then make a call without having to find someone on the server. (We talk about using those last two items in the following sections of this chapter.)

To add a name to the Personal Directory, follow these steps:

1. **In the OnLine Directory, Session List, or history folder, right-click on a person's name.**

   As shown in Figure 10-7, Internet Phone displays a pop-up menu. The menu contents vary depending on the list of names from which you're making a selection.

2. **Choose Add to Personal Directory from the pop-up menu.**

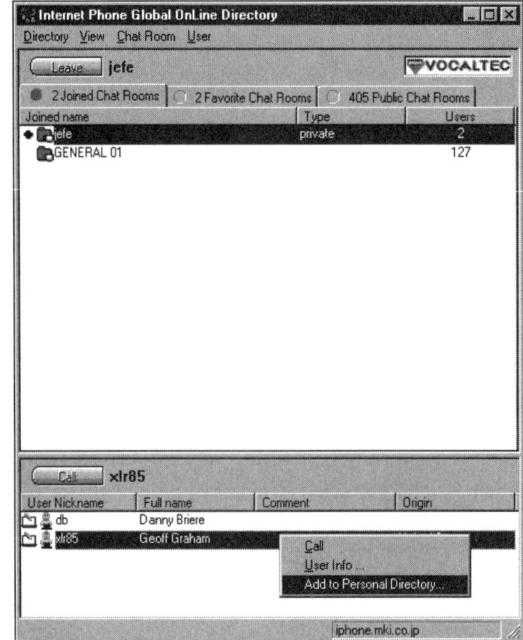

Figure 10-7: Use this pop-up menu to add someone to your Personal Directory.

Internet Phone saves the name and other user information as a file in your Personal Directory folder on your hard drive. The first 20 names you save are also added to the bottom of the Call Center menu in the main Internet Phone window.

The fact that Internet Phone saves each name as a file in a folder on your hard drive is significant, because you can then go into the folder in Windows 95 and organize the folder. For example, you can throw names you no longer want to keep in your Personal Directory into the Recycle Bin to delete, or you can add new folders within the Personal Directory to organize your phone book by categories. (You may, for example, add two folders, one for personal contacts and one for business contacts.) If you add folders within the Personal Directory folder, the names of people within each folder appear on a submenu that branches off the bottom of the Call Center menu. (The submenu branches off from the name of the folder you created on the Call menu.) So you can still see more than 20 names without actually opening the folder.

You can also add someone to your Personal Directory by dragging the person's name from the OnLine Directory, Session List, or history folder directly into the Personal Directory folder on the desktop of your computer.

To call someone whose name you've saved in your Personal Directory, simply choose that person's name from the Call Center menu. If you have more than 20 entries and you can't see the name you want, open the Personal Directory folder by clicking on the Personal Directory button on the main Internet Phone window or by choosing Call Center⇨Personal Directory from the menu bar. (When you have a chance, you may want to go into this folder and create submenus, as we discuss in the previous paragraph.) A window similar to the one shown in Figure 10-8 opens, and you can place a call simply by double-clicking on the name of the person you want to call. You can also right-click on a name to open a pop-up menu displaying other options (such as getting additional information about the person or sending a voice mail to that person).

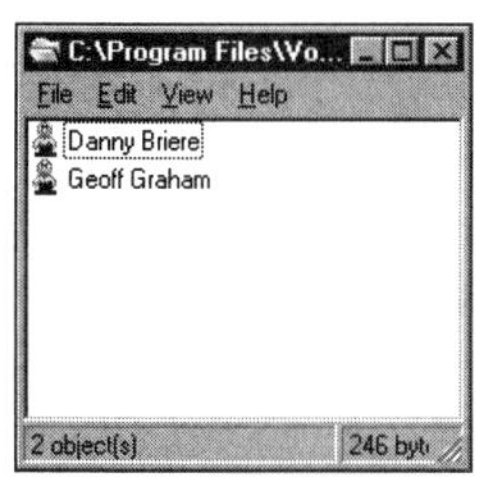

Figure 10-8: The Personal Directory window.

Tracking down someone you already talked to

Internet Phone also keeps track of everyone you've called and everyone who calls you, in two folders called the Incoming and Outgoing History folders. Well, "everyone" is a slight overstatement, but the default setting is 300. You can increase or decrease this setting (depending on your preference and hard drive space) by choosing View⇨Options from the menu bar in the main Internet Phone window. After you access the Options dialog box, click on the General icon on the left-hand side and then click on the History tab. You can click on the increment arrows next to the Internet Phone Users text box to increase or decrease the number of names you want to save. The files that Internet Phone saves are actually quite small, so unless your hard drive is very full, we see no reason to decrease the factory settings.

To call someone from either of the History folders, follow these steps:

1. **Open the History folder by choosing either Call Center⇨Outgoing History or Call Center⇨Incoming History from the menu bar.**

   The History folder opens, as shown in Figure 10-9. Like the Personal Directory window, the History folder is a standard Windows 95 window.

2. **Double-click on a name to initiate a call.**

   You can also right-click on the name to open a pop-up menu with other options, such as adding the person to your Personal Directory.

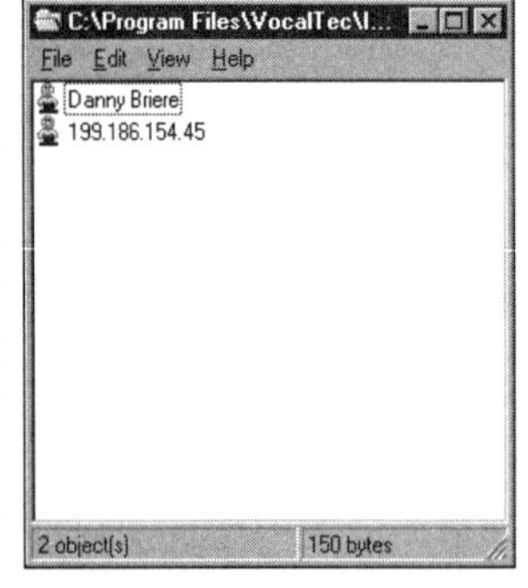

Figure 10-9: See all of your incoming or outgoing calls in the History folder.

Using the Session List

The makers of Internet Phone just don't seem to want you to lose a name — they provide yet a third way to save names of people you call (or who call you) in Internet Phone. This third way is the Session List, which we mention at the beginning of the chapter. Like the History windows, the Session List is automatically updated every time you place or receive a call. The names in this list, however, aren't saved — if you quit Internet Phone, the list is reset and starts fresh next time you log on. (That's okay, however, because the names are also saved to your History folders.)

To call someone from the Session List, simply double-click on that person's name in the list. Can't beat that for ease of use, we think.

Making a Direct Call

Internet Phone enables you to bypass the server system and call someone directly. To do this, you must know one of the two following pieces of information about the person:

- **An IP address:** Chapter 1 and several other places in this book note that an IP address is a numerical address that identifies someone's computer's location on the 'Net.
- **An Internet Phone Address:** VocalTec assigns a unique, nonchanging address (similar to an e-mail address) to each registered user. If someone shares this address with you, you can call the person directly without needing to search through the server to find that person.

In either case, the calling procedure is identical. After you log onto the 'Net and start Internet Phone, simply follow these steps:

1. **Type the IP address or Internet Phone Address into the Call text box in the Call Center panel of the main Internet Phone window.**

   The window should then look something like Figure 10-10.

2. **Click on the Call button.**

   Internet Phone attempts to connect you.

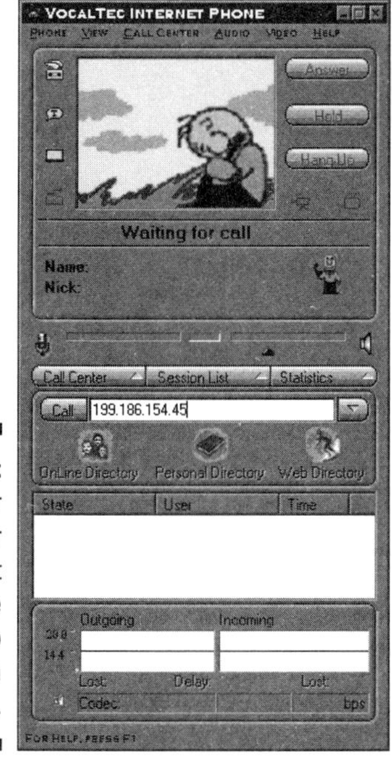

Figure 10-10: Just enter an IP or Internet Phone Address to make a direct call.

Receiving a Call

Receiving a call on Internet Phone is even easier than placing one. You can use any of the following options:

- Manual Accept
- Automatic Accept
- Do Not Disturb

Manual Accept

If you're logged into several public Chat Rooms, you probably want to use Manual Accept so that you can screen out calls before answering.

If Internet Phone detects an incoming call, you're notified by an audio notification (a ring) and a message in the Call Center reading `Call from [the person's name]`.

If you want more information before answering, click on the View Info button on the main Internet Phone window or choose Phone⇨User Info from the menu bar. You see a User Info dialog box similar to the one shown in Figure 10-11.

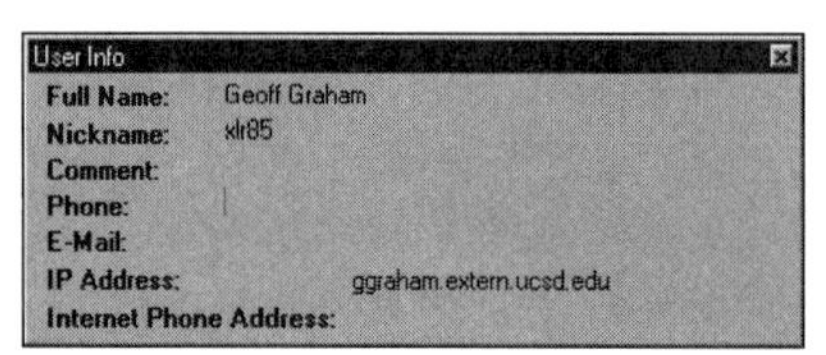

Figure 10-11: The User Info dialog box provides Caller ID information.

If you want to accept the call, click on the Answer button in the main Internet Phone window. If you don't want to accept the call, click on the Hang Up button or choose Phone⇨Hang Up from the menu bar.

Automatic Answer

If you're using only private rooms and are expecting a call, you may want to use the Automatic Answer feature. Select this feature by choosing Phone⇨Auto Answer from the menu bar in the main Internet Phone window.

This feature automatically connects incoming calls without any intervention from you. The phone "rings," Internet Phone connects the call, and the next thing you hear (with luck) is a clear "Hello" from a friend!

Blocking calls

You may decide that you don't want anyone to call you (we've all been in this mood a few times) but that you still want to log onto the server to find someone in particular. In this case, you can block all incoming calls by choosing Phone⇨Do Not Disturb from the menu bar. Any incoming calls show up in your Session List, but your Internet Phone does not ring.

What if you're already talking to someone else?

If someone calls while you're already connected to another party, the Answer button on the main Internet Phone window changes to a Call Waiting button. If you want to answer the second call, click on the button. Internet Phone places your first call on hold while you answer the second. To switch back to your original conversation, click on the caller's name in the Session List, and you can resume your call.

Other Neat Features

In addition to the features we describe in previous sections of this chapter, Internet Phone can do some other cool stuff — as if it isn't cool enough already. This new version of Internet Phone has several new features that add some multimedia functions to the basic telephone features of the program. You can perform any of the following actions:

- Use a Text Chat window to type messages back and forth with someone.
- Share pictures and drawings on the Whiteboard.
- Send voice mail messages to a person's e-mail address.
- Transfer files with the person to whom you're talking.

Using the Text Chat feature

The Text Chat feature enables you to open a window and send typed text back and forth with the person to whom you're connected. You may want to use this feature if you're having trouble understanding what the other person is saying or if you have some information to share (such as an e-mail address) that's easier to communicate in writing.

To open the Text Chat window, click on the Text Chat button in the main Internet Phone window or choose Phone⇨Text Chat from the menu bar. A window similar to the one shown in Figure 10-12 opens on-screen. All you need to do is type into the bottom of that window and press the Enter key to send your text. As Ross Perot says, “It's that simple!”

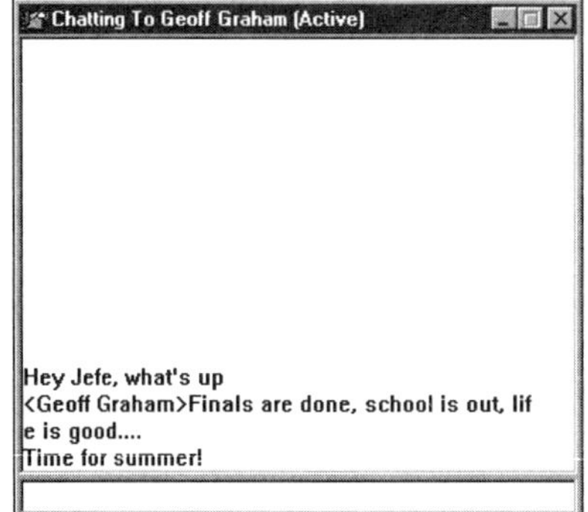

Figure 10-12: Sometimes typing is better than talking!

Using the Whiteboard

The *Whiteboard* is an interactive, electronic version of the whiteboard or chalkboard that you have in your office or classroom. You can paste files (graphics files or images of a spreadsheet, for example) onto the Whiteboard; draw on it by using a set of standard drawing tools; and then mark up, highlight, erase, or otherwise deface what you put onto the Whiteboard. As you perform these actions, the person on the other end of your call sees everything you do and can add pictures, drawings, comments, and so on — and those additions show up on your screen for your amusement.

We often find that the first thing anyone does with a whiteboard (ourselves included) is to draw a smiley face on it, similar to the one shown in Figure 10-13. You can really do a whole lot more, however. The first time one of us used the Internet Phone Whiteboard, for example, he was talking to a nice fellow from New Zealand who was living in Hong Kong. The gentleman from New Zealand/Hong Kong opened the Whiteboard and pasted onto it some photos he had on his hard drive. We were able to share a great view of the spectacular Hong Kong skyline (taken from an apartment across the way in Macau) and some photos of New Zealand's capital city, Wellington. Pretty neat stuff — although, to be honest, we did indeed start off by drawing smiley faces before the photos started coming across the 'Net.

Whiteboards are powerful tools, but in limited bandwidth situations, sending a large file across the 'Net while talking can result in poor audio quality and a general slowing down of your computer. The effect lasts for only a little while, but be forewarned that such a transmission can have a drastic effect on your audio quality.

Figure 10-13: The Whiteboard.

Sending a voice mail message

Another neat new feature of Internet Phone is its capability to send voice mail messages to people's e-mailboxes. You don't even need to limit yourself to other Internet Phone users; you can send a voice mail message to just about anybody who has an Internet e-mail account. The message is received as an *attachment* (a file that accompanies an e-mail) that can be played back by Internet Phone or by a free voice mail player that VocalTec distributes on its Web site (at `http://www.vocaltec.com`, in case you forgot), so your friends who don't use Internet Phone can simply download the player and listen to your messages.

To send a voice mail message, perform the following steps:

1. **In the main Internet Phone window, click on the Voice Mail button (the one that looks like a picture of a cassette tape with wings) or choose Phone⇨Voice Mail from the menu bar to open the Voice Mail dialog box (see Figure 10-14).**
2. **Address the message by typing an e-mail address in the To text box.**

   If you're sending the message to someone who is in your Personal Directory, click on the To button, and a standard Windows file dialog box opens. Double-click on the name of the person to whom you want to send the voice mail.

   Note: Sometimes people don't put their e-mail address into their personal information when they configure Internet Phone. In such cases, you can't add those addresses to a voice mail message sent from your Personal Directory — you must enter the e-mail address manually.

Figure 10-14: Send a voice mail over the 'Net.

3. **To send a copy of the voice mail message to someone else, enter the other person's name in the cc text box.**

   You can also add people from your Personal Directory by clicking on the cc button and following the same procedure as you did in Step 2.

4. **To begin recording your voice mail message, click on the Record button (the one with the red dot, on the far left-hand side of the toolbar in the dialog box).**
5. **Speak into the microphone for as long as two minutes.**
6. **After you finish your message, click on the Stop button (the one with the black square, on the far right-hand side of the toolbar) to stop recording.**

   You can play your message back to yourself, if you want, by clicking on the Play button (the second from left, with the single arrow on it). If you don't like your message, you can repeat Steps 4 through 6 until you are happy with the message.

7. **If you want to add text, do so in the Text Message text box.**

   Text appears in the body of the e-mail message that you send.

8. **Click on the Send button to forward your mail to your selected recipients.**

As we note at the beginning of this section, if you receive a voice mail message, the message appears as an attachment to your e-mail message. To listen to the voice message, you simply extract the file and then double-click on the Files icon. (Exactly how you extract the file depends on how your e-mail program works, but most have a simple button you can click on to do the job.) Internet Phone starts up and plays back the message.

Transferring files

Got a neat new shareware program you want to show a friend? You don't need to remember the Web address where you downloaded the program; just call your friend on Internet Phone and transfer the file. Doing so is really quite easy; just follow these steps:

1. **While you are connected with your friend, open the File Transfer tool by clicking on its icon (a folder with an arrow above it) in the main Internet Phone window or by choosing Phone⇨File Transfer from the menu bar.**

   The Select File for Transfer dialog box (a standard Windows file dialog box) opens, as shown in Figure 10-15.

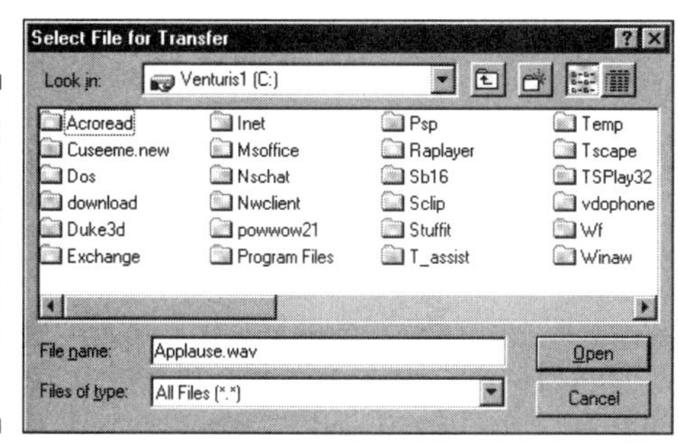

Figure 10-15: Browse through your hard drive to find a file to transfer.

2. **Use the dialog box to browse through your hard drive and find the file you want to send. After you find the file, click on its name to select it and then click on OK.**

   The Transfer File dialog box opens, as shown in Figure 10-16.

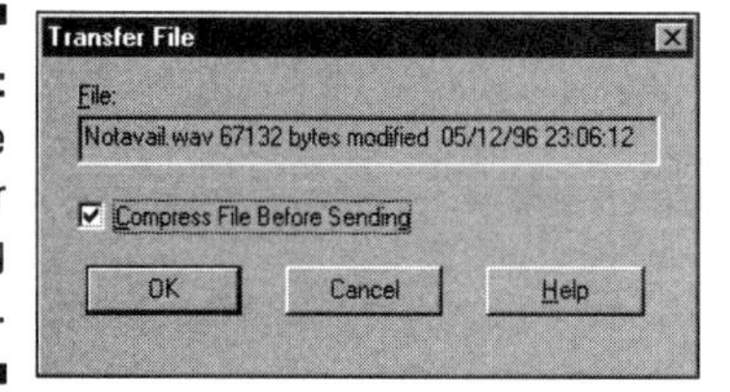

Figure 10-16: The Transfer File dialog box.

   If you are connected to more than one person (that is, if you have someone on hold), Internet Phone opens the Participant's dialog box rather than the File Transfer dialog box. You must select the person for whom the file is intended by clicking on that person's name in the Participants dialog box.

3. **Leave the Compress File Before Sending check box selected.**

   Compressing the file helps limit the effect that its transmission has on your voice connection, because compression reduces the amount of bandwidth you need for transferring the file.

4. **Click on OK.**

Internet Phone transmits the file transmitted to the other party.

Say Goodnight, Gracie: Quitting Internet Phone

Done for the night and want to quit? This process is a really easy one; just follow these steps:

1. **Choose Phone⇨Exit from the menu bar in the main Internet Phone window.**
2. **If you want, close your ISP 'Net connection.**

Like all 'Net telephone products, Internet Phone works only if you're logged onto the 'Net. So after you quit the program, you can't receive any calls or any notification of calls that are attempted while you're offline. If you have a dedicated 'Net connection, you may want to keep Internet Phone running in the background the whole time your computer is running. This strategy works best if you joined only a private Chat Room; if you joined several public Chat Rooms, you can expect your Internet Phone to be ringing with calls from people you don't want to bother you — for example, the Howard Stern imitator who called while one of the authors was writing this section!

Internet Phone for the Macintosh

We hate to leave our fellow Mac users out in the dark, so this section offers a quick rundown on the Mac version of Internet Phone. In general, it's a lot like the Windows version we describe in the preceding pages, so if you skim through that information, you'll basically be ready to use the Mac version. We want to tell you up front, however, that the Mac version differs from its Windows counterpart in the following ways:

- It has no videoconferencing capability.
- You can't share files by file transfer.
- You can't conduct a text chat.

- It doesn't have a whiteboard.
- You can't put calls on hold

The Mac version gives you a basic, audio-only Internet telephony program that lets you talk to other Mac users and to users of the Windows version of the program. So even though you lose a lot of functionality, you still have a phone that gives you access to a whole bunch of people.

The Mac version of Internet Phone (shown in Figure 10-17) has an interface that's very similar to the Windows version. As you can see, you have that cute little animated assistant fellow feeding you information about your call status, a Call Center (with the Session List integrated into it), an Audio Setup section to adjust your microphone and speaker levels, and a Statistics Window. The interface also includes a Caller Information section, which shows you the same information as the About User dialog box in the Windows version.

Just like the Windows version, Internet Phone for the Mac uses the Global OnLine Directory. (You can see the OnLine Directory in Figure 10-18.) Like in the Windows version, you can join as many as 10 Chat Rooms at once, and you can create your own private or public Chat Rooms. You can also filter the Chat Room list if you find some room titles offensive or want to keep your kids out of the adult areas.

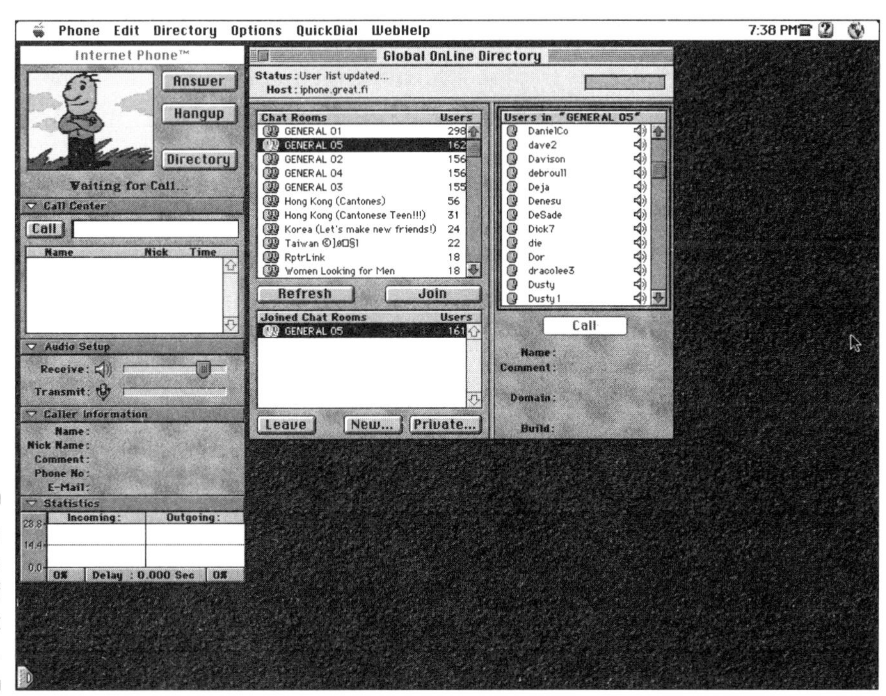

Figure 10-17: The Mac version of Internet Phone.

All of the other controls for Internet Phone for the Macintosh reside up in the Mac OS menu bar. We skip the Apple menu and the Edit menu (which are just standard Mac OS menus) and tell you about the others:

- **Phone:** Lets you answer and hang up on calls, and opens the Global OnLine Directory.
- **Directory:** Contains controls for connecting and leaving the OnLine Directory server, entering and leaving chat rooms, placing calls and filtering chat.
- **Options:** Lets you set up your preferences for Internet Phone for the Macintosh.
- **QuickDial:** Opens your speed dial phone book of saved names of people you want to call.
- **WebHelp:** Provides access to the online help features of Internet Phone.

Making calls with Internet Phone for the Mac is extremely easy (just like all Mac programs, right?). You can place a call in one of two ways:

- From the OnLine Directory
- Directly, by entering an IP address in the Call Center

Calling from the OnLine Directory

Calling someone from the OnLine Directory involves just a few steps. Here's what you do:

1. **Start your Internet connection.**

   Internet Phone won't start if you don't have a network connection.

2. **Open Internet Phone by double clicking on its icon.**

   With the default setup for Internet Phone, the OnLine Directory opens automatically.

3. **If you've set your preferences so that the OnLine Directory doesn't open automatically, choose Phone⇨Directory from the menu bar (or press Command+O).**

4. **Join the chat rooms of your choice by double-clicking on their names in the Chat Room listing.**

   The names of joined Chat Rooms appear in the Joined Chat Room listing at the bottom of the OnLine Directory, while the nicknames of people in the selected Chat Room appear in the Users in Chat Room listing.

5. **Search through the Chat Room for the person you want to call and then double-click on that person's nickname.**

   That's all you have to do. Internet Phone figures out the person's Internet address and places the call for you.

6. **When you finish the call, just click on the Hang Up button on the main Internet Phone window to terminate your connection.**

One neat feature of the Mac Internet Phone is the capability to search for a particular user on the OnLine Directory without knowing which Chat Room that person is in. Here's all you have to do:

1. **Choose Directory⇨Find (Command+F) from the menu bar.**

   As shown in Figure 10-18, Internet Phone displays a Find dialog box.

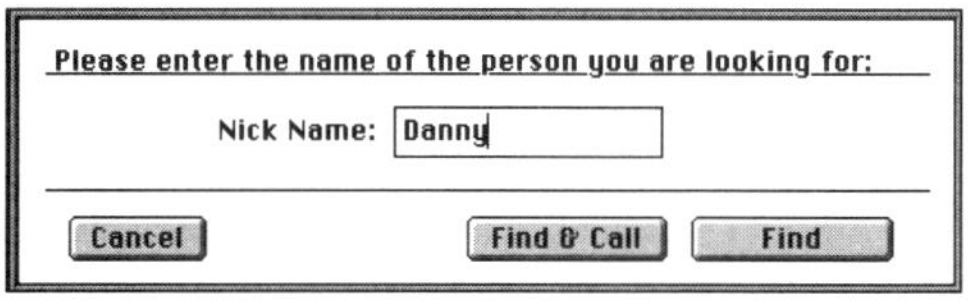

Figure 10-18: You can search the OnLine Directory for a particular person.

2. **Enter the nickname of the person you are looking for in the text field.**

3. **Click on the Find & Call button if you want Internet Phone to look for the other party and place the call for you. If you want to make sure you've got the right person, click on the Find button instead.**

   Internet Phone searches the entire OnLine Directory for the person you want, and it returns a dialog box with the Caller Information for that person.

4. **If you've found the right person, click on the Call button in the dialog box, and Internet Phone places the call for you. If you don't find the person you want, click on Cancel to close the dialog box.**

Pretty slick, huh? Windows users have to search manually through the OnLine Directory to find someone, or they must go to the Web Directory.

The virtues of being direct

Placing a direct call is even easier than using the OnLine Directory. Well, it's easier if you know the IP address or host name for the person you want to call. But if you have that little nugget of information, calling directly is a piece of cake:

1. **In the main Internet Phone window, type the host name or IP address of the person you want to call in the Call Center Edit Box.**
2. **Click on the Call button.**

Internet Phone rings the remote party's phone and connects the call.

Using the QuickDial feature

You find your Internet Phone for the Mac phone book in the QuickDial menu. Adding a user to the phone book is simple. The only caveat is that you must be connected to other people to add them to your QuickDial list. Here's how you do it:

1. **Connect to the other user (via the OnLine Directory or a direct call).**
2. **Choose QuickDial⇨Add User (Command+S) from the menu bar.**

Internet Phone adds the user's nickname to the menu. To call this person in the future, simply select the person's name from the QuickDial menu. That's all it takes.

Getting back to work

When the time comes to finish your Internet Phone session and get back to work (or dinner, or bed, or wherever), just choose Phone⇨Quit (Command+Q) from the menu bar. Now you can get back to that Maelstrom game.

Chapter 11

Using Intel Internet Video Phone on Your PC

In This Chapter

- Setting up Internet Video Phone for the first time
- Making that first call
- Using Directory Assistance and speed dial
- Making and receiving calls

Internet Video Phone is a snap to use. It has an interface that's easy to use and almost foolproof. Unlike some of the other, more business-oriented, phones out there, Internet Video Phone doesn't have a lot of collaboration tools like whiteboards or file transfer capabilities. Instead, it's designed so you can easily get online and make a video call to someone else.

Internet Video Phone uses the same sort of system as Microsoft NetMeeting to locate and connect users to each other: A ULS (*User Location Service*) identifies each user's call information (such as name and current IP address), and Web-page based directories provide listings of everyone who's online. In fact, Internet Video Phone and NetMeeting share many of the same directories — which isn't surprising, because Intel and Microsoft worked together on developing the first H.323 Internet telephony programs and shared technology to make it happen. (Microsoft gave Intel its T.120 technology, and in return, Intel gave Microsoft its H.323.)

A Quick Look around Internet Video Phone

The first thing you should do with your copy of Internet Video Phone is to take a quick look at all the buttons and screens available to you and get a general idea of what they do. Figure 11-1 shows the main screen of Internet Video Phone.

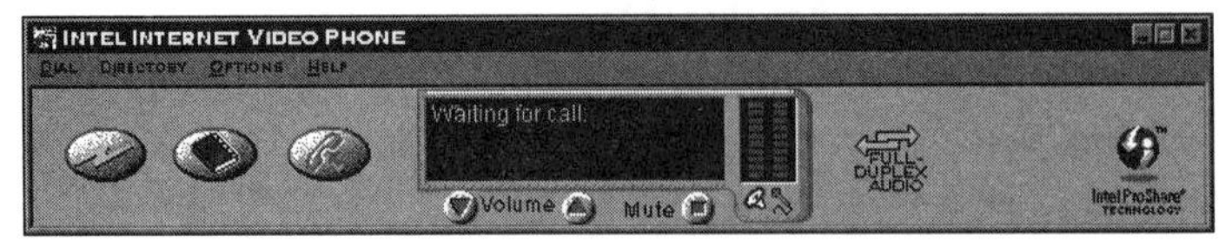

Figure 11-1: The Internet Video Phone main window.

Across the top of Internet Video Phone, you have a menu bar with the following menus:

- **Dial:** This menu contains commands for making calls, answering calls, and exiting calls, as well as for accessing the Call Preferences dialog box.
- **Directory:** This menu offers you easy access to all the online Web directories used by Internet Video Phone. The directory you choose from this menu opens in Internet Explorer.
- **Options:** The Options menu lets you control various settings for Internet Video Phone and allows you to find out more information about your current call.
- **Help:** As you may have guessed from the name, the Help menu provides access to the help system, which is Web based and opens in Internet Explorer.

Below the menu bar, Internet Video Phone has four buttons, a status window, and audio controls. From left to right, here's a brief overview of each item:

- **Speed Dial button:** This button opens the Speed Dial dialog box for dialing frequently called numbers.
- **Directory button:** Clicking on this button opens your preferred Web phone book in Internet Explorer.
- **Answer/Hang Up button:** This button toggles between Answer and Hang Up, depending on your call status.
- **The Status window:** This window displays the current status of Internet Video Phone — for example, waiting for a call or in a call.
- **Volume controls:** The up arrow increases your speaker or headphone volume, and the down arrow decreases the volume. You also have a mute button, which you can click on to mute your microphone; click on this button a second time to turn the mike back on.
- **Energy meters:** These meters give you a graphical representation of your audio signal levels.
- **Talk button:** If you are using half-duplex audio, you must press the Talk button to send audio. With full-duplex audio, the talk button is replaced by an icon that says Full-Duplex audio, because you don't need to push to talk. (That's what you see in Figure 11-1.)

The other main interface elements you see when you use Internet Video Phone are the video windows. These windows appear only when Internet Video Phone makes a video connection. If you have video, you see the My Video window, and if the other party in the call has video as well, you see a Guest Video window. These two windows are pretty much the same, with some minor differences, and they look like the window in Figure 11-2.

Figure 11-2: A video window.

The last bit of interface in Internet Video Phone is the Intel Connection Advisor, shown in Figure 11-3. The connection advisor lets you know how your CPU and Internet connection are handling your call — and it gives you advice if it finds a problem. To open the Intel Connection Advisor, choose Options⇨Check Intel Connection Advisor from the menu bar or click on its icon in the Windows 95 taskbar tray. Like the video windows, the Connection Advisor can be opened only during a call.

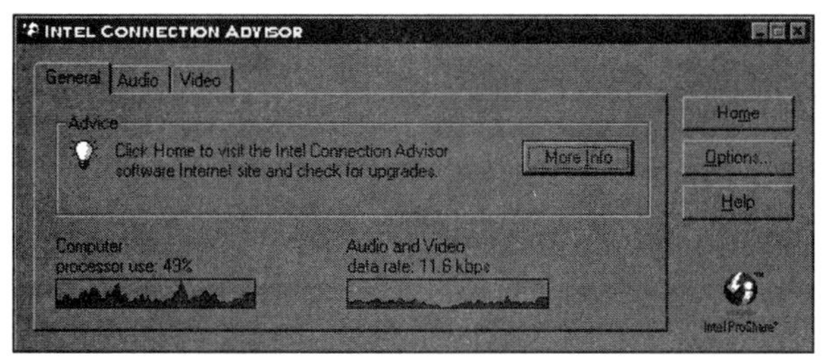

Figure 11-3: The Connection Advisor can tell you what's happening during a call.

Configuring Internet Video Phone

Like most of the major Internet telephony and video products, the installer for Internet Video Phone automatically sends you through a Configuration Wizard that asks you questions and configures your system based on your answers. This easy process collects the necessary information about your audio and video settings and basically gets you all set up to make that first call. If you ever change your audio or video hardware or drivers and wish to reset your configuration, you can reopen the Configuration Wizard by choosing Start⇨Programs⇨Intel Internet Phone⇨Intel Internet Video Phone Configuration Wizard from the Windows 95 start menu.

Before you begin making calls, you may want to check out four preferences dialog boxes in Internet Video Phone:

- Call Preferences
- Preferences
- My Listing
- H.323 Firewall Settings

Most users never venture into the H.323 Firewall Settings dialog box unless curiosity leads them there. If you are behind a firewall and need to use this dialog box, get very cozy with your network administrators because they can tell you what to fill in here — we can't, because the correct information depends on your individual firewall.

Establishing calling preferences

As a first step to getting ready for a call, you should make sure that you have your Call Preferences set up properly. This task takes about 20 seconds to complete — here's how you do it:

1. **Start Internet Video Phone by double-clicking on its Shortcut icon on the desktop or by choosing Start⇨Programs⇨Intel Internet Phone⇨Intel Internet Video Phone from the start menu.**

   You must have your Internet connection active to open Internet Video Phone, even if you don't plan to make a call — otherwise it just won't open.

2. **Choose Dial⇨Call Preferences from the menu bar.**

   The Call Preferences dialog box opens (see Figure 11-4).

3. **In the Type of call area, click on the radio button corresponding to your preference.**

   If you have video capabilities, you probably want to click on the Send and receive video radio button. That's what we always select.

4. **In the Resize My Video window section, click on the radio button for the size of video window you want to send to the other party.**

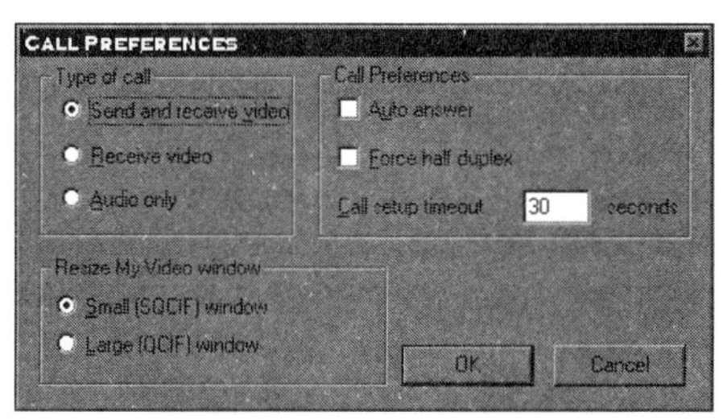

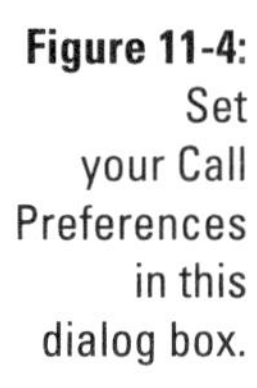

Figure 11-4: Set your Call Preferences in this dialog box.

Keep in mind that a larger video window requires more screen space and CPU usage to reproduce on the other end of your call. You may want to stick with the Small window size until you know that a larger size works properly on the other party's system.

5. **In the Call Preferences section, click to select the Auto answer check box if you want Internet Video Phone to answer incoming calls automatically.**

   If you select this option, make sure to turn the camera away if you decide to change clothes while Internet Video Phone is running. If someone calls, your video pops up automatically on the other person's screen!

6. **Click to select the Force half duplex check box if you want to make your audio work in half duplex.**

   Note: Intel advises that you use this setting if you have audio troubles because of a low bandwidth connection. We recommend that you start off with this box unselected, and resort to it if you have to — full duplex is much nicer than half!

7. **Leave the Call setup timeout setting at the default 30 seconds.**

   If you seem to be having trouble connecting to people, you may want to increase this time to ensure that latency on the 'Net isn't keeping you from hooking up with someone.

8. **Click on the OK button to save any changes you have made, and close the dialog box.**

Getting listed

After you establish your calling preferences (as we detail in the preceding section of this chapter), you need to ensure that you are properly listed on a ULS system. Actually, you do so during the installation procedure (the Wizard prompts you to enter this information), but why not take 10 seconds to make sure that you're all set? Here's what you do:

1. **Open the My Listing Information dialog box by choosing Options➪My Listing Information from the menu bar.**

   The dialog box looks like the one you see in Figure 11-5.

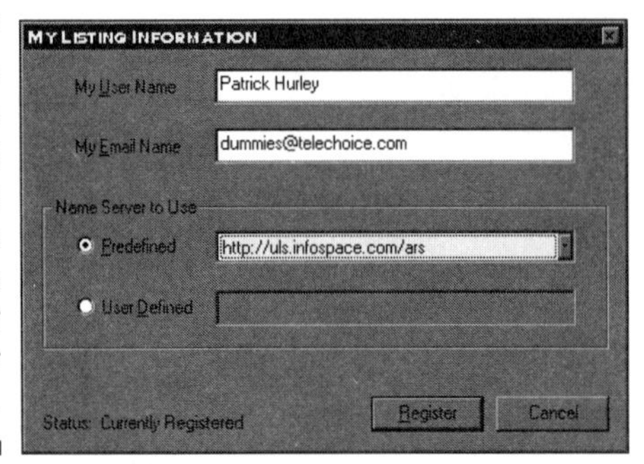

Figure 11-5: Use this dialog box to ensure that you are listed on the ULS server of your choice.

2. **Make sure that your name and e-mail address appear in the appropriate text fields in the dialog box — if they don't, enter them in the correct fields.**

3. **In the Name Server to Use section of the dialog box, click on the radio button next to Predefined.**

   Note: You can select your Name Server (ULS server) from a predefined list, or you can enter a user-defined ULS. The only case we can imagine in which you may want to enter your own ULS would be in an intranet environment, where your company runs its own server. Otherwise, use one of the predefined ULS servers.

4. **Select the server you wish to use from the drop-down list.**

   Remember that the ULS Name Server you log onto will be the Web page on which you are listed for other people to find and call — so if you have some friends who also use Internet Video Phone, you may want to choose the same server as them.

   Note: For some of the Name Servers in the predefined list, you must register yourself separately before logging onto them. If you get an error message saying that you can't log onto the directory you've chosen, choose Help➪Contents from the menu bar. The Internet Video Phone help Web page has links to the registration Web pages for the various directory services.

5. **Click on the Register button to log onto the ULS Name Server, and you're done.**

Setting your general preferences

Before you start making calls, you need to set your general preferences. Follow these steps:

1. **Choose Options⇨Preferences from the menu bar.**

   The Preferences dialog box opens, as shown in Figure 11-6.

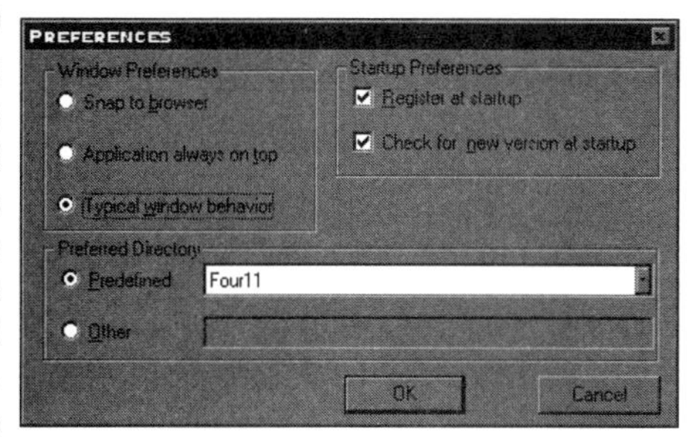

Figure 11-6: Make Internet Video Phone behave like you want it to.

2. **In the Window Preferences section, click on a radio button to select your personal preference.**

   Your selection is truly a matter of what you are most comfortable with and how large your display is. Try all three options and see which one you like best:

   - **Snap to browser:** Makes Internet Video Phone attach itself to the bottom of your browser window when you open the Web phone book or the online help.
   - **Application always on top:** Keeps Internet Video Phone on top of other open windows at all times.
   - **Typical window behavior:** Makes Internet Video Phone behave like any other Windows application.

3. **Decide whether to select the two check boxes in the Startup Preferences section of the dialog box.**

   Normally, you keep both of these check boxes selected:

   - **Register at startup:** Automatically registers you with your preferred ULS directory server when you start the program.
   - **Check for new version at startup:** Does just what it says it does — it checks in with Intel over the 'Net and ensures that you have the latest version of Internet Video Phone.

4. **Use the Preferred Directory section of the dialog box to specify which directory service opens when you click on the Directory button in Internet Video Phone.**

   (You can always view any of the directories by selecting the one you want from the Directory menu.) You must first decide whether you want to use a predefined directory (your best choice at this point in time) or whether you want to enter a directory manually. Click on the radio button next to Predefined or Other to make this choice. We strongly recommend that you choose Predefined unless a particular directory that you know about and want to use doesn't appear in the list of predefined ones (for example, if you are on an intranet and your company runs its own internal directory).

5. **Select the directory of your choice from the drop-down list next to Predefined or enter the entire URL of your directory next to Other.**

6. **Click on the OK button to save your changes and close the dialog box.**

That wasn't too hard. After you've set all of your preferences, you're ready to make that first call.

Making Your First Call

Internet Video Phone offers you several options for placing a call. You can connect to someone in any of the following ways:

- By selecting the person from a Web phone book or directory service
- By using an IP address
- By selecting a name that you have saved in your Speed Dial list

Calling from a Web directory

The most common method for placing a call is by calling from a Web directory. Just follow these steps:

1. **Establish an Internet connection (if you use a dial-up ISP connection).**
2. **Start Internet Video Phone.**
3. **Click on the Directory button.**

Your Web browser opens to your preferred directory — for example, the Four11 directory. All of them are pretty similar, but they have some minor differences.

Note: If you want to look for someone to call in a directory other than your preferred one, simply select that particular directory from the Directory menu on the menu bar, and it opens in your browser window.

4. **Using the links and the search capabilities of the directory window in your browser, locate the entry for the person you want to call.**

   As shown in Figure 11-7, your directory window may include links for showing all online users of Internet Video Phone and all users of H.323-compliant Internet telephony programs, as well as tools for searching the directory to find a particular person.

   Again, we must warn you that all directories have slightly different options, but anyone who has even a modest ability to navigate the World Wide Web should have no difficulty figuring out how to complete this step.

5. **After you find the person you wish to call, either as the result of a search or by browsing through the list of online users, simply click on the link that says Intel Internet Video Phone, as shown in Figure 11-8.**

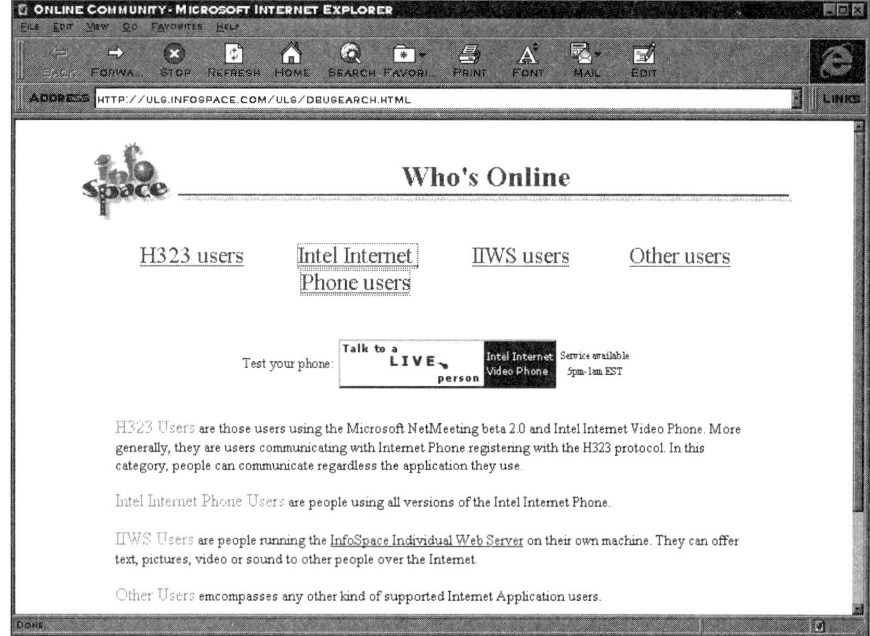

Figure 11-7: Infospace's Internet Video Phone Web directory.

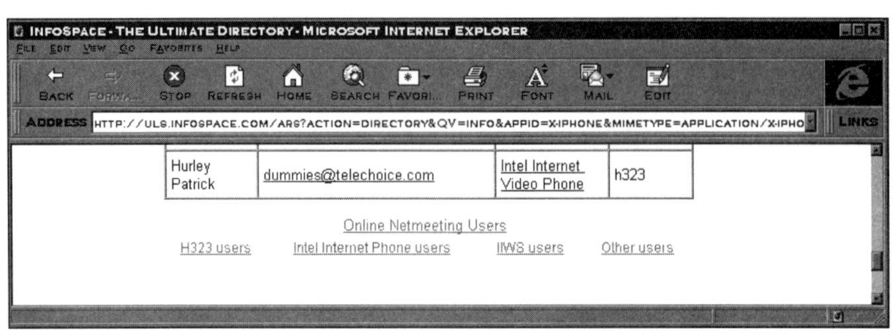

Figure 11-8: Click on Intel Internet Video Phone to make a connection.

That's all you have to do. The directory service automatically figures out the IP addresses for you and the person you are calling and feeds Internet Video Phone the right information to place the call.

6. **You're talking now (well, as long as the other person accepts your call)!**

   To adjust the volume of your speaker or headphones, just click on the up- or down-volume arrows in the Volume section of the main Internet Video Phone window. You can also mute your microphone by clicking on the Mute button. Click on it again to reactivate your microphone.

If you, or the other party, have video and you've elected to send and receive it (you make this choice in the Call Preferences dialog box), the My Video and Guest Video windows pop up and display the video stream.

After you finish speaking, click on the Hangup button or choose Dial⇨Hangup to disconnect the call.

Video fun

During a videoconference, you can do a few things to the video windows in Internet Video Phone. For example, you can pause your own video — in the My Video window — and leave a still image up on the screen. This is a great feature to know about in case you need to blow your nose during a call. Here's how you do it (pause your video, that is):

1. **In the My Video Window, choose Pause⇨Stop sending video.**

   This command freezes your video.

2. **When you are ready to send again, choose Pause⇨Start sending video.**

If the other party is sending video, you can access some controls in the Guest Video window to adjust the quality of the incoming video:

1. **In the Guest Video window, choose Options➪Guest Quality Controls.**

   The Guest Video: Options dialog box opens, as shown in Figure 11-9.

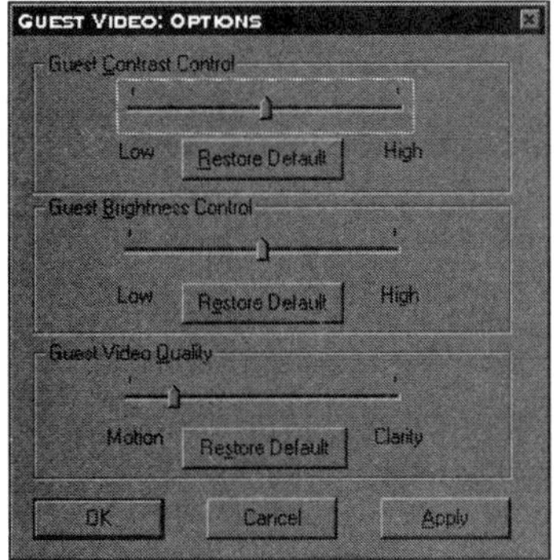

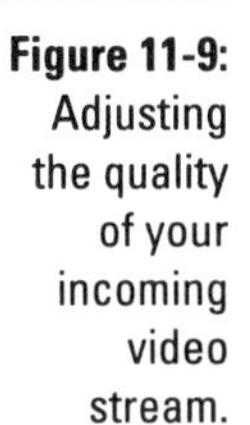

Figure 11-9: Adjusting the quality of your incoming video stream.

2. **Use your mouse to adjust the sliders left or right, according to your preferences.**

   You can adjust the following settings:

   - **Guest Contrast Control:** Adjusts the level of contrast — just like a contrast control on your TV.
   - **Guest Brightness Control:** Dims or brightens the incoming video picture.
   - **Guest Video Quality:** Allows you to choose between a clearer picture (the clarity end of the scale) or one that moves like actual video instead of a series of stills (the Motion end of the scale).

3. **If you want to see the results of your changes, click on the Apply button.**

4. **When you are satisfied with the video, click on OK to save your changes and close the dialog box.**

Placing a direct call

If you know a person's IP address or host name, you can skip the ULS server system and call the person directly. This is, as you might imagine, a very simple process:

1. **Choose Dial⇨Direct Dial from the menu bar.**

   A dialog box opens, as shown in Figure 11-10.

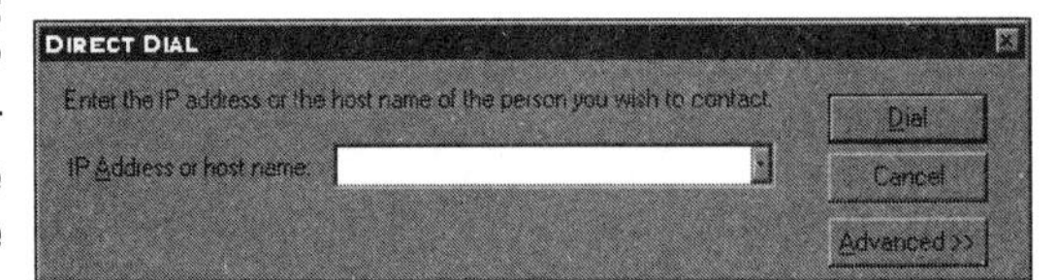

Figure 11-10: Enter an IP address or host name here to make a direct call.

2. **Type the IP address or host name in the text field.**
3. **Click on the Dial button to place the call.**

That's all you need to do. Internet Video Phone connects to the IP address and notifies the other person that you are calling.

Remember to say "Hello"

If you receive an incoming call, your speakers ring, and notification of the incoming call appears in the Status window of the main Internet Video Phone window, as shown in Figure 11-11.

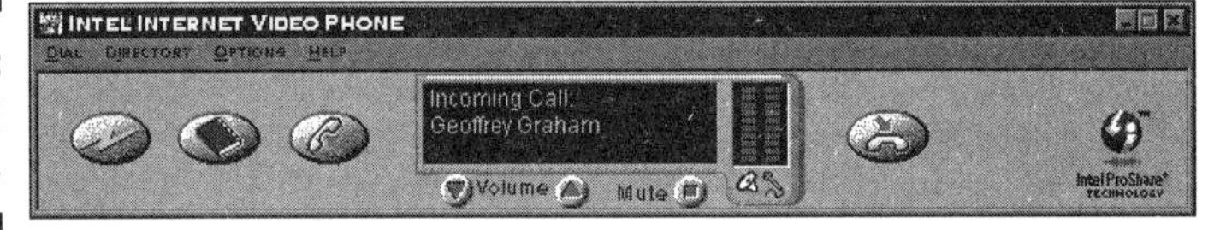

Figure 11-11: An incoming call.

To answer, simply click on the Answer button (or press Ctrl+A on your keyboard). That's it!

If you don't feel like talking, choose Dial⇨Reject Incoming Call or press Ctrl+J on your keyboard.

Star 69 (Not the REM song)

You know how you can get a service from your telephone company that lets you punch in a few numbers (usually *69) and return calls that you miss? You can do the same thing with Internet Video Phone, and it doesn't cost you a penny (unlike those 75-cents-a-pop services the telephone company sells you). Here's all you have to do:

1. **Choose Dial⇨Unanswered Calls from the menu bar.**

   The Unanswered Calls dialog box pops open on your screen, as shown in Figure 11-12. Any calls you've missed show up in this listing (with a bunch of information like name and time of the call).

2. **Click on the name of the person you want to call and click on the Dial button to call back.**

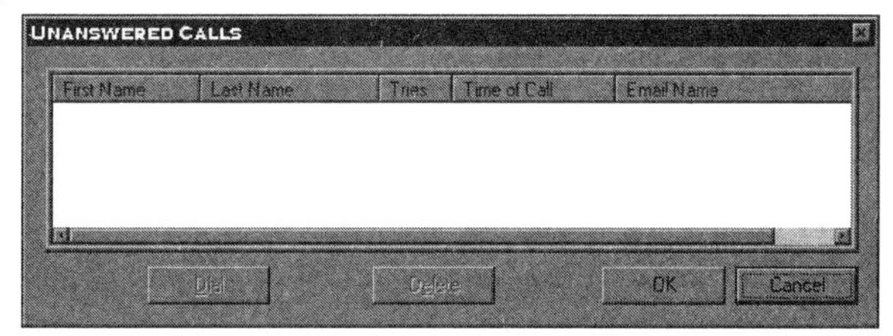

Figure 11-12: See a listing of calls you've missed in this dialog box.

If you have your phone set to the Auto Answer mode (set in the Call Preferences dialog box, which we discuss earlier in this chapter), the Unanswered Calls recorder won't work (because you won't have any unanswered calls!).

Speed dialing your way to happiness

Rather than search through directory listings for the folks you call most, why not add them to a Speed Dial listing, and call them with a click (or two) of the mouse? Internet Video Phone lets you do this quite easily.

Entries are made to speed dial after you complete a call. You have to set up your preference ahead of time — so you may as well do that now:

1. **Click on the Speed Dial button on the main Internet Video Phone Window.**

   The Speed Dial dialog box opens, as shown in Figure 11-13. On the right-hand side of the dialog box, you see a section titled Add to List After Call.

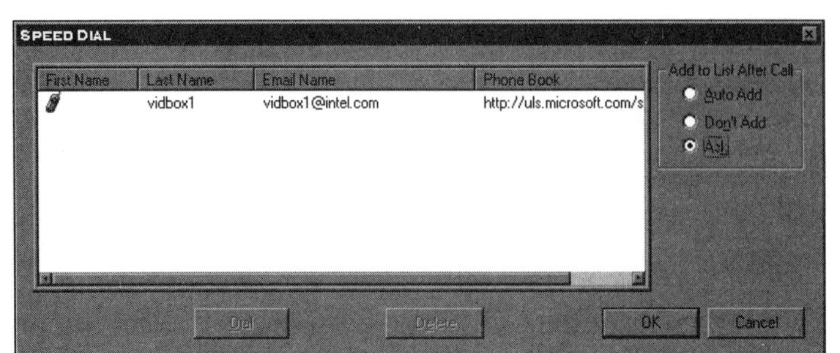

Figure 11-13: The Speed Dial dialog box.

2. **Click on the radio button corresponding to the method you want to use for building your Speed Dial list.**

   You can choose any of three options:

   - **Auto Add:** Internet Video Phone automatically places everyone you talk with in the Speed Dial listing.
   - **Don't Add:** No one gets placed in Speed Dial.
   - **Ask:** Internet Video Phone lets you decide after each call whether to add the caller to the Speed Dial listing. (After you complete each call, you get a small dialog box asking whether you want to add that person to your listing.)

3. **Click on OK to save your preferences.**

When you want to call someone who's in the Speed Dial listing, just follow these steps:

1. **Click on the Speed Dial button to open the Speed Dial dialog box.**
2. **Click on the first name of the person you wish to call.**

   You can sort your Speed Dial list by clicking on one of the headings (First Name, Last Name, Email Name, and Phone Book).

3. **Click on the Dial button to initiate the call.**

Ciao!

Tired of talking? Time to go to bed? Well, then you should quit Intel Internet Video Phone. It's easy — one of those one-step procedures that make us so happy. Just choose Dial⇨Exit from the menu bar.

Good night!

Chapter 12

Using Netscape Conference

In This Chapter

- It's no longer CoolTalk, but it's still cool
- Conferencing can be fun
- Find and be found
- Collaboration isn't a bad word

Conference (previous versions were called CoolTalk) is part of Netscape's forthcoming Communicator software package. Communicator, in case you don't know, is the next generation of Netscape client software, and it includes a new version of the Navigator Web browser, an e-mail program, a newsreader, and whole bunch of other stuff — including Conference. The version that we talk about in this chapter is the fourth preview release (basically, a beta version with a different name), but we've been told that the interface and basic functions will be the same whenever the final version is released. So don't expect to see anything very different — but don't be surprised if you find some minor variations in the version you download from Netscape. By the way, you can find Communicator on the Netscape Web site at `http://www.netscape.com`.

Conference is tightly integrated into the Communicator package. Although the main interface appears as a separate program and window on your computer's desktop, Conference shares an address book with the Communicator e-mail program and uses Navigator to display the Phonebook listing of currently online users. This tight integration reflects a growing trend: As Internet telephony becomes more common, you'll see Internet phones integrate even more seamlessly with other programs — as Internet phone plug-ins for Web browsers, perhaps, or integrated into online networked games.

Conference is an H.323-compatible Internet telephony program, so you can use it to call users of other H.323 products, like NetMeeting or Intel Internet Video Phone. However, Communicator doesn't use the User Locator Service (ULS) directory system that these phones use — it uses Netscape's own variation on the theme called DLS, or Dynamic Lookup Service — so you have to use a third-party service such as Four11 to make calls to other H.323 programs.

Netscape offers Windows 95 and Macintosh versions of Communicator, but — and this is the standard "but" we always put in after a statement like this — the Windows version is much farther along in development and features than the Mac version. For that reason, we focus primarily on the Windows version in this chapter.

Snooping Around

Look around the Conference interface and get acquainted with the features and functions we discuss in this chapter. Figure 12-1 shows the main Conference window.

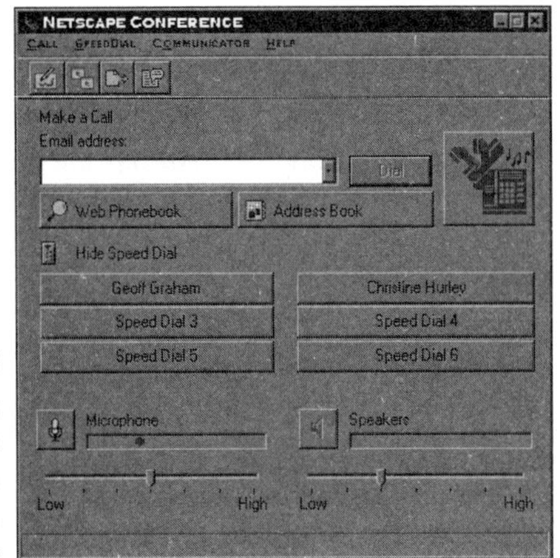

Figure 12-1: Conference's main window.

Across the top of the main window, Conference has the following menus:

- **Call:** This menu allows you to place and end calls, control your preferences, and exit the Conference program.
- **SpeedDial:** This menu controls the six speed dial entries — you can make calls from here or edit the entries.
- **Communicator:** A common menu in all Communicator modules, this menu allows you to open or switch to other elements of Communicator.
- **Help:** This menu opens the help system and activates the Setup Wizard.

Beneath the menus, the Conference toolbar contains four buttons that perform the following functions (from left to right):

- **Whiteboard:** Opens the Whiteboard
- **Collaborative Browsing:** Opens the controls for collaborative browsing

- **File Exchange:** Opens the controls for the file transfer function
- **Text Chat:** Opens the Text Chat window

Farther down on the main window, Conference offers the following features:

- The Call window, a text field for entering the e-mail address of the person you wish to call. This field also offers a drop-down list from which you can select a previously entered address.
- A Dial button for initiating a call to the selected e-mail address. While you are connected on a call, this button changes to become the Hang Up button — with an obvious function!
- Buttons for opening the Web Phonebook and the Address Book.
- An information button for displaying the business card of the person with whom you are talking. (If you aren't in a call, it displays version information for Conference.)
- Six Speed Dial buttons, and a button to toggle the display of the Speed Dial buttons. (The Speed Dial buttons are displayed in Figure 12-1.)
- Microphone and speaker displays and controls.

Setting Up Conference on Your PC

Setting up Conference is very easy — the program uses a Setup Wizard to guide you through the process.

In case you've never seen wizards before, they are a common Windows feature designed to lead users, step by step, through a complex process. Wizards are most often seen in Microsoft Office applications such as Excel (for example, the Chart Wizard).

Off to see the wizard

When you run Conference for the first time, the program automatically invokes the Setup Wizard, which leads you through the setup process. You can rerun the wizard at any time, and you may want to do so if your system configuration changes (for example, if you get a new sound card or a new modem). To start the wizard manually, follow these steps:

1. **Choose Help⇨Setup Wizard from the menu bar.**

   The Setup Wizard dialog box opens on your desktop (see Figure 12-2).

2. **Follow the steps as they appear, clicking on the Next button to move from one step to the next.**

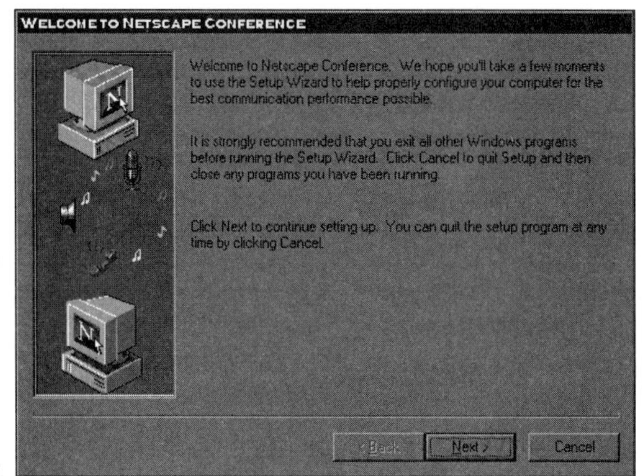

Figure 12-2: The Setup Wizard can be your friend.

These steps are so easy that we're not going to waste ink describing each one.

3. **After you confirm all of your settings in the wizard and reach the end, click on the Finish button.**

   You're all set.

Options, options, options

You don't have to run through the Setup Wizard to change some of your most often used settings if you don't want to. You can change your Business Card, Network, and Audio settings at any time by choosing Call⇨Preferences from the menu bar. For example, if you want to change the DLS directory server and the Web phonebook you use, just follow these steps:

1. **Choose Call⇨Preferences from the menu bar.**

   The Preferences dialog box opens. (See Figure 12-3.)

2. **Click on the Network tab.**

3. **In the DLS server text box, use the drop-down list to select the name of the server you want to use.**

   For example, to use Four11's DLS server, select `netdls.four11.com`. This is the only server currently available, although others are likely to emerge in the future. If you know of another server that's not available in the list of servers, you can simply enter its address.

4. **In the Phonebook URL box, select the Web phonebook that you want to have open in Netscape Navigator when you click on the Web Phonebook button.**

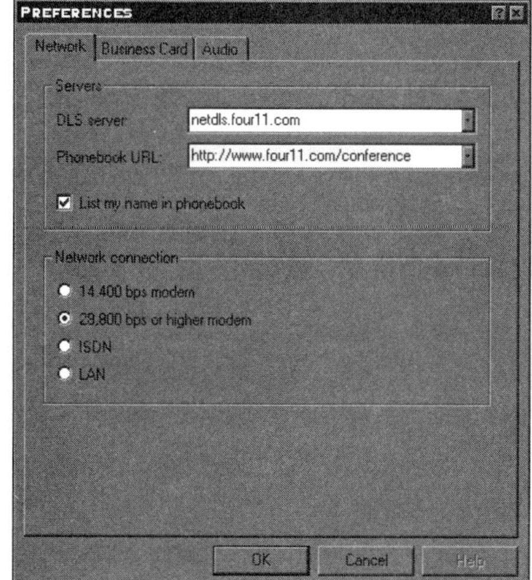

Figure 12-3: Change your options in the Preferences dialog box.

The default setting is `http://www.four11.com/conference`, but you should be able to use other Web phonebooks as they become available for Conference. If you want your name to appear on the Web phonebook (so that people can see you and call), be sure that the List my name in phonebook check box is selected.

5. **If you want to log on privately, click on the List my name in phonebook check box to deselect it.**

   With this setting, people who know your e-mail address can still call you via the DLS server, but you won't be listed in the Web phonebook.

   The Setup Wizard should have already taken care of your Network connection settings, but if it hasn't, you can select the appropriate radio button, depending on the speed of your 'Net connection.

6. **Click on OK, and you're finished.**

Before you start making and receiving calls, you should set up one last thing. Conference gives you three options for handling incoming calls. In the Call menu, you can choose from the following options (a check mark appears next to the option you choose):

- **Always Prompt:** All incoming calls cause an invitation dialog box to pop up and ask you whether you wish to receive a call.
- **Auto Answer:** Conference automatically answers all incoming calls, without any intervention on your part.
- **Do Not Disturb:** Conference blocks all incoming calls.

After you choose the desired options, you can start conferencing!

Conferencing Know-How

Conference enables you to make calls in any of the following ways:

- *Directly,* by entering an address (host name or IP address) into the Direct Call dialog box
- *From the DLS directory server system,* using an e-mail address
- *From Netscape Navigator,* using the Web Phonebook on Four11's Web site

All calls, except those to an IP address, use the DLS directory server system to determine the other party's location on the 'Net; IP address calls go directly to the other party. In practice, you never really notice the difference between the two approaches in terms of actually making a call.

Conference automatically logs onto the server at startup, as long as you specified the correct DLS server (like `dls.four11.com`) on the Network tab of the Preferences dialog box.

Calling using the DLS system

Here's how you make a call using the DLS directory:

1. **Log onto your ISP (if you don't have a dedicated connection).**

2. **Start Conference from the Start menu or by double-clicking on its icon.**

   If you already have Netscape Communicator running, you can launch Conference by choosing Communicator⇨Conference from the menu bar.

3. **Enter the e-mail address of the person you want to call in the Call window of the main Conference window.**

   You can also select an address from the drop-down list of recently called addresses.

4. **Click on the Dial button on the main window to initiate the call.**

   You can also choose Call⇨Dial from the menu bar or just press the Enter key to start dialing.

   A Pending Invitation dialog box pops up to display your conference request status. After Conference makes the connection and the other party accepts your call, you see the person's picture in the Conference Info window (if they've included a picture in the business card), and you're ready to speak.

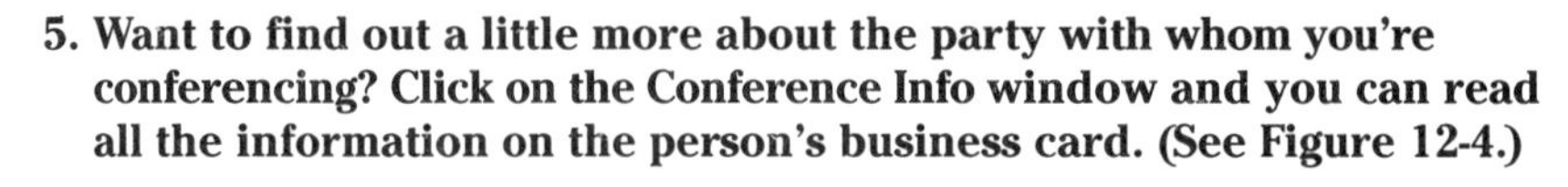

5. **Want to find out a little more about the party with whom you're conferencing? Click on the Conference Info window and you can read all the information on the person's business card. (See Figure 12-4.)**

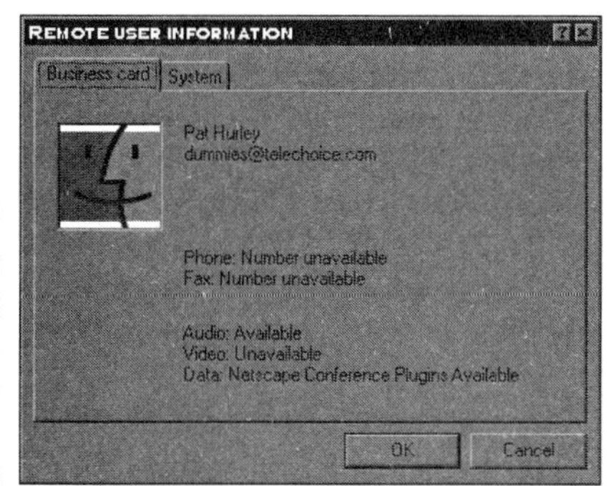

Figure 12-4: Looking at someone's business card.

6. **After you're done speaking, choose Call⇨Hang Up from the menu bar or click on the Hang Up button on the main window.**

Adjusting your call quality

You can adjust the audio quality of your Conference call in these ways:

- By adjusting the silence level
- By adjusting the echo suppression level
- By adjusting the playback and microphone levels

The silence level setting serves the same purpose as the microphone activation level settings that we show you in Chapters 9, 10, and 11 — it establishes the point at which Conference distinguishes your voice from background noise and begins sending out audio. To set it, follow these steps:

1. **In the Conference main window, click on the Microphone button and then place your mouse pointer on the blue diamond in the Microphone Display window.**

   You should see a horizontal arrow and the words `Silence Level`. (See Figure 12-5.) If you're not speaking, any green bars showing should be to the left of the red pointers.

2. **Speak into your microphone and make sure that your normal voice level is to the right of the pointers (as indicated by the green bars).**

   In general, you want to set the silence level as far to the left as possible without letting background noise trigger your audio.

If you have a half-duplex sound card, this setting is crucial. If you set the silence level too low, your sound card constantly sends out audio data and is never free to receive incoming audio.

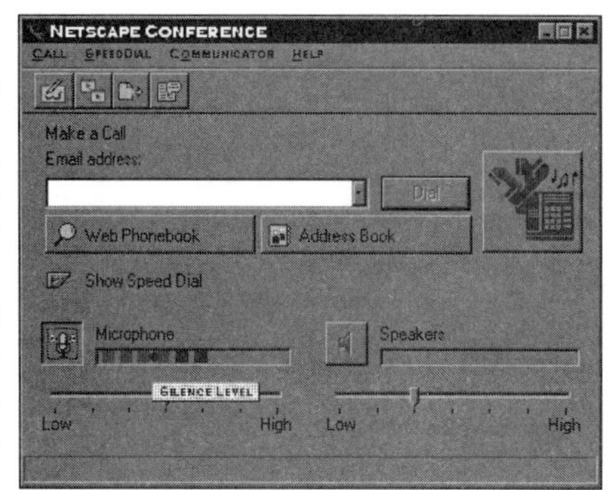

Figure 12-5: When you talk, the microphone level should appear to the right of the silence level indicator.

If you use a full-duplex sound card, you want to pay attention to the echo suppression level adjustment. Echo suppression basically keeps the incoming sound of the other party from being sent back through your microphone and causing feedback. If you're silently listening to the other party, you don't want that person's voice to activate your audio and get sent back to the other party.

You set the echo suppression level in the Preferences dialog box:

1. **Open the dialog box by choosing Options⇨Preferences.**
2. **Click on the Audio tab of the dialog box to see the audio preferences shown in Figure 12-6.**
3. **Click on the arrow and select your preferred echo suppression level from the drop-down list.**

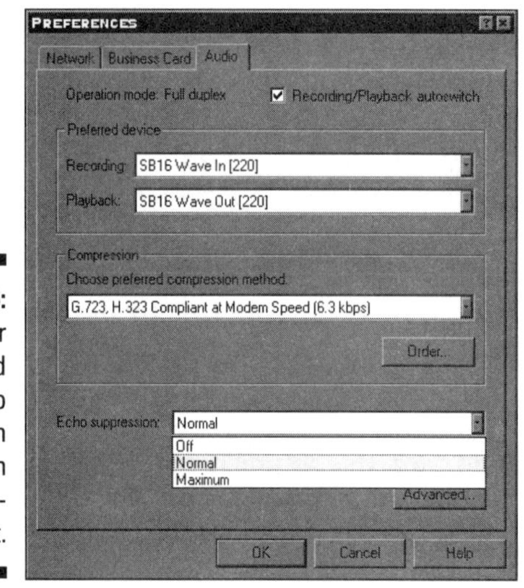

Figure 12-6: Select your preferred Echo suppression level from the drop-down list.

You have the following choices:

- **Off:** Use this mode if you have a telephone-style handset or use headphones.
- **Normal:** Use this mode for general speakerphone-style conferencing.
- **Maximum:** Choose this mode if you experience feedback problems. For example, if you use a laptop, the speakers may be especially close to the microphone.

4. **Click on the OK button to save your changes and close the Preferences dialog box.**

Reach out and touch someone on the Web

The Conference Phonebook on Four11's Web site offers perhaps the easiest way to find and connect to other users of Conference. Using the Conference Phonebook, you can search for the person you're trying to find and then click on his or her name to connect.

Here's how it works:

1. **With Conference running and your Internet connection active, click on the Web Phonebook button.**

   The Conference Phonebook page opens in a Navigator Window, as shown in Figure 12-7.

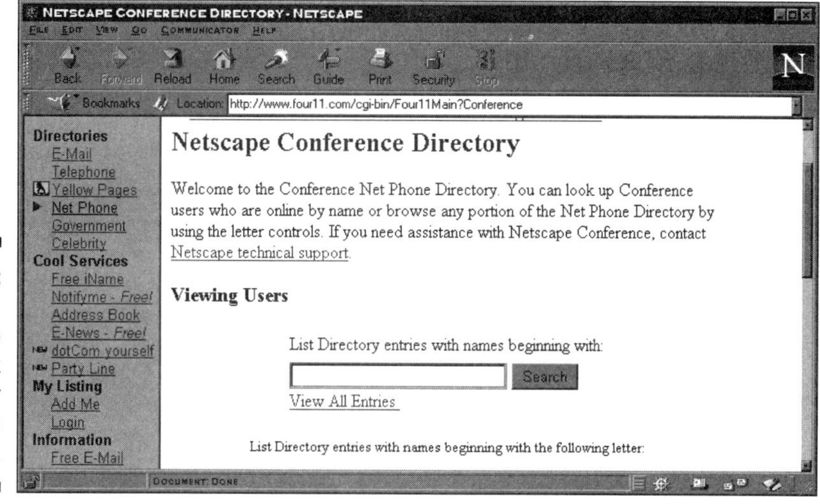

Figure 12-7: The Four11 Web Phonebook for Conference users.

2. **If you want to search for a particular person, enter that person's name in the text field on the Web page and then click on the Search button.**

   You can also view all users or groups of users alphabetically by clicking on the other links.

   If Four11 has a listing for the person you want, and that person is online, your party's name and e-mail address appear in a new frame on the Navigator page.

3. **To initiate a call to this person, click on the Conference link next to the person's e-mail address.**

   Conference automatically places your call.

In fact, you don't even have to be running Conference ahead of time. Just aim Navigator at the Four11 Conference directory manually and click on a link to call someone; Conference automatically launches and connects the call.

Adding someone to your Conference Address Book

Conference shares an address book with the rest of Communicator — so the same address book that you use to save e-mail addresses for example, can be used to place calls with Conference.

To put someone in your address book:

1. **Click on the Address Book button on the main Conference window.**

   The Address Book window opens, as shown in Figure 12-8.

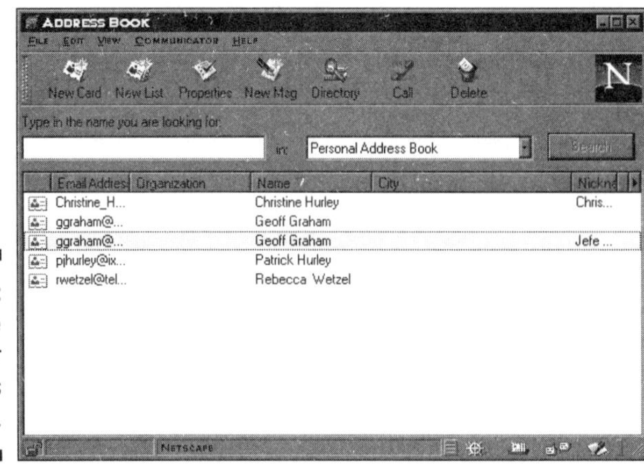

Figure 12-8: The Communicator Address Book.

2. **Click on the New Card button in the Address Book window.**

   The New Card dialog box opens.

3. **Click on the Name tab and fill in all of the information you have on this person.**

   Make sure you include at least the First Name, Last Name, and Email Address fields if you plan to use a DLS server to call this person.

4. **Click on the Netscape Conference tab (shown in Figure 12-9).**

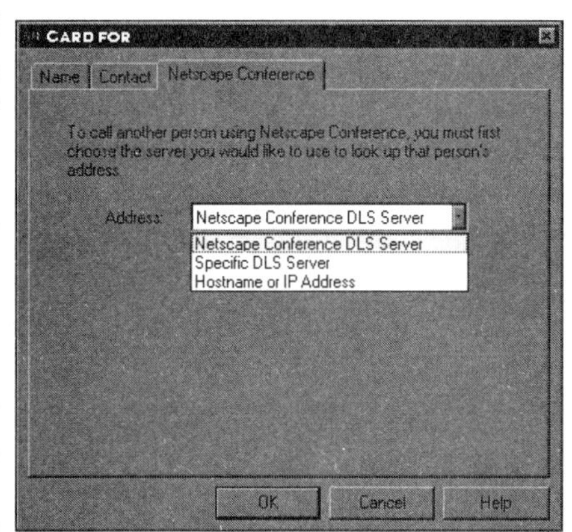

Figure 12-9: Select your Conference calling preferences for an individual you're adding to your address book.

5. **In the Address window, select Netscape Conference DLS Server.**

   Note: You can also select and enter data for other DLS servers or put a fixed IP address here, if desired.

6. **Click on OK to close the dialog box and save your changes.**

Calling from the Address Book

Making a call from the Address Book is easy; just follow these steps:

1. **With Conference running, click on the Address Book button on the main window.**

   The Address Book opens.

2. **Click on the address book entry for the person you want to call.**

3. **Click on the Call button.**

Using Speed Dial

Conference lets you save six Speed Dial settings for the folks you call most often. To add a user to your Speed Dial list:

1. **Right-click on one of the Speed Dial buttons.**

   The Speed Dial Edit dialog box opens, as shown in Figure 12-10.

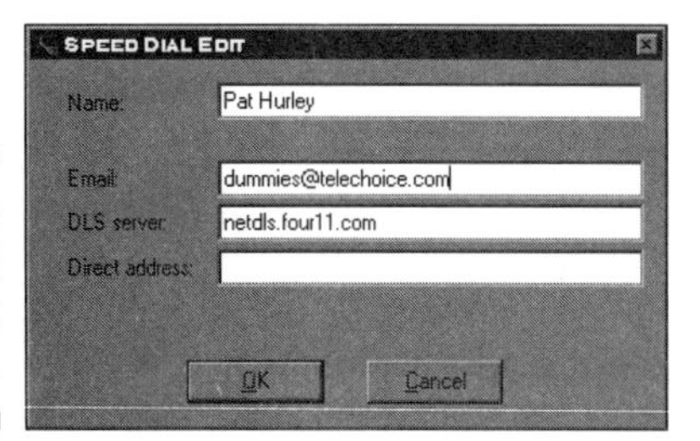

Figure 12-10: The Speed Dial Edit dialog box.

2. **Enter the person's name and e-mail address in the appropriate text fields in the dialog box.**

   If the person has a fixed IP address, you can enter it here as well.

   Note: Conference automatically fills in the DLS server for you. As other DLS servers become available, you can enter a preferred server address in this box.

3. **Click on OK to save your settings and close the dialog box.**

   The person's name appears on the specified Speed Dial button in the main Conference window — replacing the generic "Speed Dial" label that the button displayed. Sure is a lot easier than trying to write on that little cardboard thing on your regular phone!

To place a call to someone on your Speed Dial list, just click on the Speed Dial button for that person.

During a call with someone, you can give that person some of your valuable Speed Dial real estate. Choose Speed Dial⇨Speed Dial *X*⇨Replace (where *X* represents the Speed Dial number you want to use). This command automatically fills in the Speed Dial information based on that person's business card and DLS server information.

Receiving a call

Unless you selected Always or Never Accept Invitation during the set-up process, Conference displays a pop-up dialog box to notify you of incoming calls. (See Figure 12-11.) To answer the call, click on Accept Call and then click on the OK button. If you want to decline the call, click on Reject Call and then click on the OK button.

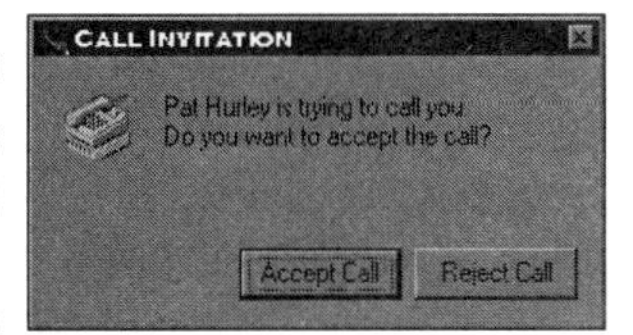

Figure 12-11: An incoming call!

Sharing Is Caring

The interactive, multimedia part of Conference comes into play with collaborative features such as the Whiteboard, File Exchange, and Collaborative Browsing. Sometimes talking just isn't enough — and Internet telephony really is better than regular telephony. Try doing these things with your Mickey Mouse phone!

Better than a chalkboard, but no erasers to clap

The Whiteboard enables you to perform any of the following actions:

- Share images and graphics files.
- Paste screen shots from your desktop or open applications.
- Type comments onto the Whiteboard.
- Draw or illustrate on the Whiteboard by using standard drawing tools (such as those found in a paint application).
- Selectively delete or erase sections of the Whiteboard.
- Use a pointer to highlight specific sections of the Whiteboard.
- Reconcile the contents of the Whiteboard in case of transmission errors.

You activate the Whiteboard by clicking on the Whiteboard icon on the toolbar. You can open the Whiteboard at any time. (Although opening the Whiteboard when you're not connected generally doesn't do you much good, you may want to put something on the board ahead of time while you're preparing for a call.) The blank Whiteboard looks something like the example in Figure 12-12. The Whiteboard has two distinct layers or levels:

Figure 12-12: The Whiteboard.

- **The Image Layer:** Contains images captured from the screen or loaded from files on disk
- **The Markup Layer:** Contains drawings, text, and Clipboard items pasted onto the Whiteboard

Why two layers? The Whiteboard enables you to erase or clear items on the Markup Layer without changing the underlying images on the Image Layer — so you and your friends can look at a screen shot of the same document, add comments or draw faces on it (or whatever you normally do), and, if things get too crowded, erase only the comments, leaving the original image untouched. Of course, you can erase everything, if you like, and start fresh.

The Whiteboard's user interface consists of the *main canvas,* where the Image and Markup Layers live. To the left of the canvas is the drawing toolbox, which should be familiar to anyone who has ever used a drawing program like Paint. The toolbox contains the following tool sections:

- **Tools:** Contains a pencil; an eraser; line, circle, and rectangle drawing tools; a pointer; and a text tool
- **Width:** Controls the size of items drawn onto the Markup Layer
- **Fill:** Determines whether items drawn on the Markup Layer are solid or only partially filled in
- **Color:** Of course, adds color

The Whiteboard also contains the following six menus in its menu bar:

- **File:** Used to open files to place on the Image Layer and to save Whiteboards (for review later)
- **Edit:** Controls Whiteboard synchronization and clearing and pasting of objects from the Clipboard
- **View:** Enables you to zoom in and out (magnify and minimize)
- **Capture:** Enables you to capture items from the desktop and paste them on the Image Layer
- **Options:** Controls the behavior of the Whiteboard — such as whether it opens automatically, what the eraser erases, and so on
- **Help:** Gives you . . . Heeeellllllllppppppp!!!!

Using the Whiteboard is pretty straightforward, and getting into it in detail is beyond the scope of this book, but here are a few points to remember:

- Sending large files back and forth on the Whiteboard can eat up a great deal of bandwidth, so don't expect your call quality to remain quite as good while you send a full-sized screen shot or a big graphics file — especially if you're on a slow modem connection.
- Before you use the eraser tool, check the Options menu and see what you're set to erase; you can either erase the Markup Layer only or the whole thing (Markup and Image Layers).
- If you have a transmission error, or if for any reason your Whiteboard file becomes corrupted, choose View⇨Refresh from the menu bar.

We also want to point out the difference between a whiteboard and a document-sharing groupware application. A whiteboard enables you to share an image of a screen only from within an application, whereas groupware actually enables you to share the same document — that is, make edits and comments right into the spreadsheet or document that you're sharing. Those kind of applications are the next step beyond whiteboarding — and are quite a bit more expensive and complicated.

Using Text Chat

As do several other programs we discuss in this book, Conference includes a Text Chat function. Conference's Text Chat is a little more sophisticated than most because the feature also enables you to perform the following actions:

- Import text files from your computer directly into the Text Chat and then post them.
- Save a log of your Text Chat onto your hard drive.

To invoke the Text Chat feature during a conference, click on the Text Chat button on the toolbar. The Text Chat window opens, looking similar to the one shown in Figure 12-13.

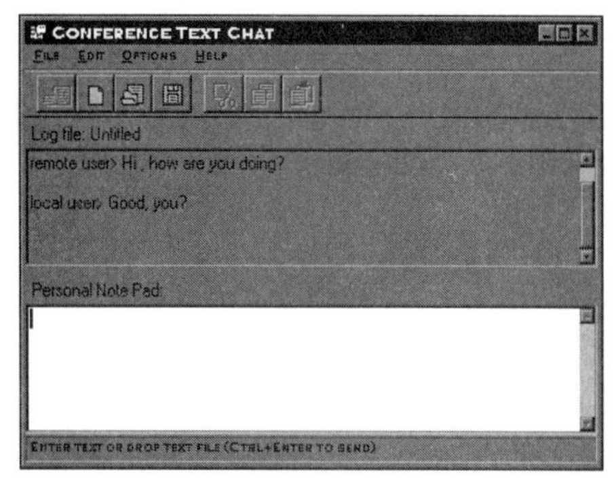

Figure 12-13: Have a text chat conversation using the Text Chat window.

This feature actually behaves a little bit like a text editor program; you can cut, copy, and paste by using the Edit menu or the toolbar at the top of the window, and you can even undo edits that you've made (also from the Edit menu). To import a text file into the Conference Text Chat window, follow these steps:

1. **Open the Conference Text Chat window by clicking on the Text Chat button on the toolbar.**
2. **Choose File⇨Include from the menu bar (or click on the folder-shaped Include button).**

   Conference opens a standard Windows file dialog box.
3. **Select the text file from your hard drive and click on Open.**

   The file appears in the bottom half of the Conference Text Chat window (your Personal Note Pad).

To send the text file or anything that you have typed in the Personal Note Pad, just press Ctrl+Enter or click on the Send icon on the toolbar.

Browsing the Web with someone else

A particularly neat feature of Conference is its Collaborative Browsing feature. Collaborative Browsing let you lead — or be led — on a tour of the Web while you are in a conference with someone. With this feature, you don't have to memorize or cut and paste the URL of your favorite new Web site to let your friend know about it — just take your friend on a guided tour. Here's how you do it:

1. **Connect with someone using any connection method.**
2. **Click on the Collaborative Browsing button on the main Conference window.**

   The Collaborative Browsing dialog box opens, as shown in Figure 12-14.

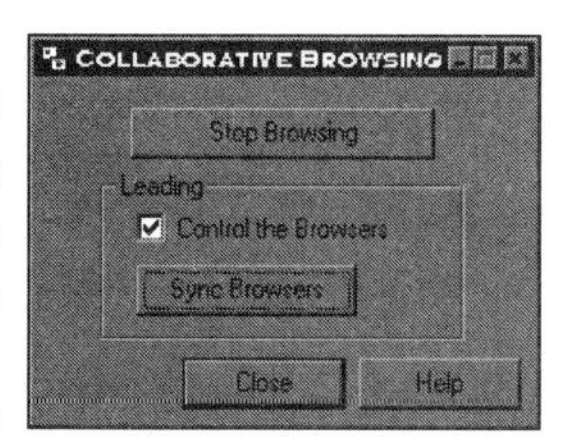

Figure 12-14: Browse the Web in sync!

3. **Click on the Start Browsing button.**

   Navigator opens, if you don't already have it open.

   A dialog box opens on the other party's screen to invite that person to join you in Collaborative Browsing. If the other user agrees, the Navigator window also opens on that party's screen.

4. **Point your browser to the URL you want to share and click on the Sync Browsers button.**

   This button automatically sends the other user to the same URL.

5. **When you finish browsing, click on the Stop Browsing button in the Collaborative Browsing dialog box.**

The person who initiates the Collaborative Browsing session leads the session. If you want to switch so the other person leads, the current leader must click on the Control the Browsers check box in the Collaborative Browsing dialog box to deselect it, and the new leader must click on it to select it.

Sharing files

Part of the purpose of an Internet telephony program such as Conference is collaboration. The Whiteboard is one example of this capability; file transfer is another. Using the file transfer tools, you can send any type of file to the person with whom you are conferencing. You can send applications, documents, graphics — whatever you have on your hard drive, in fact. Here's how you send a file:

1. **Start a conference with the person to whom you wish to transfer files.**
2. **Click on the File Exchange button on the main Conference window.**

   The Conference File Exchange window opens, as shown in Figure 12-15.

3. **Choose File⇨Add To Send List from the menu bar in the Conference File Exchange window.**

   This command opens a standard Windows 95 file dialog box.

   You can also click on the Open button in the Conference File Exchange window (the middle of the three buttons right below the menu bar).

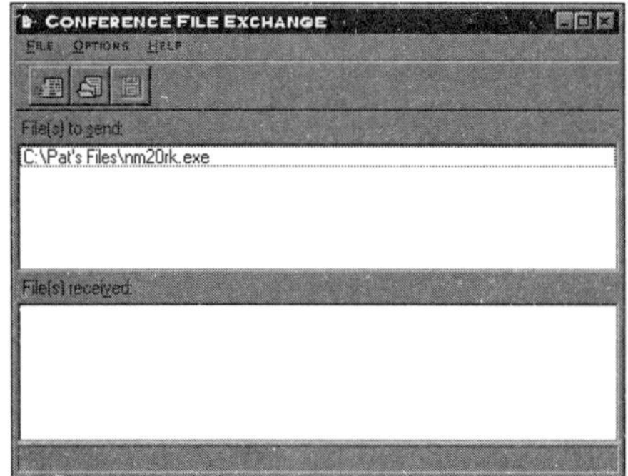

Figure 12-15: The Conference File Exchange window.

4. **Browse through your hard drive(s) until you find the file you wish to transfer, select it, and click on Open.**

   Conference adds the file to your File(s) to send listing in the Conference File Exchange window.

5. **Choose File⇨Send to send the file to the other user.**

   You can also click on the Send button on the button bar or press Ctrl+Enter.

   Note: By default, Conference compresses files before transferring them. Although we don't recommend it, you can turn off the compression by choosing Options⇨Compress from the menu bar.

6. **After you send the file, choose File⇨Close to close the Conference File Exchange window.**

Received files are listed in the File(s) received listing of the Conference File Exchange window. By default, this window pops up when you receive a file (you can turn off this feature by choosing Options⇨Pop Up On Receive, if you wish). To save received files to your hard drive:

1. **Click on the name of the file in the File(s) received listing.**
2. **Choose File⇨Save from the menu bar (you can also click on the Save button in the button bar or press Ctrl+S).**

   Conference opens a standard Windows Save File dialog box.

3. **Use this dialog box to specify a destination for the file and then click on Save.**
4. **After you save your received files, choose File⇨Close to close the Conference File Exchange window.**

Talking to the answering machine

As do an increasing number of Internet telephony applications, Conference has a built-in voice mail system that lets you send messages to people who aren't online or can't receive your call. The voice mail system sends your message as a specially encoded e-mail message which can only be played back from within Netscape Communicator.

Because Conference's voice mail system uses e-mail to send voice mail messages, it doesn't automatically prompt you to send voice mail if a call you make to an IP address or host name doesn't connect — because you haven't provided an e-mail address while making the call. However, you can send voice mail to someone directly, if you know that person's e-mail address, regardless of how you try to call the person (we tell you how in a moment).

Leaving a message

If you call someone who's online and running Conference but away from the computer, you can leave that person a message. Leaving a message is easy to do, really, because you don't need to do much except talk. Just follow these steps:

1. **Make your call normally.**

   If the call fails to connect, a dialog box pops up to ask if you would like to send a voice mail message.

2. **Click on the Yes button and the Netscape Voice Mail dialog box opens, as shown in Figure 12-16.**
3. **Click on the red circular Record button and speak your message into the microphone.**
4. **Click on the square gray Stop button after you are done.**

   If you want to check your message, click on the triangular gray Play button to play it back and review. If you don't like the message, simply click on the Record button again and rerecord your message.

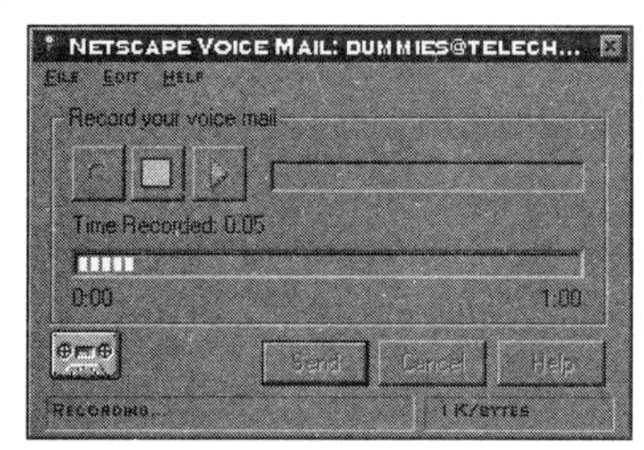

Figure 12-16: The Voice Mail dialog box — click on the red button to record a message.

5. **When you're happy with your message, click on the Send button.**

   Communicator's Composition program opens, as shown in Figure 12-17.

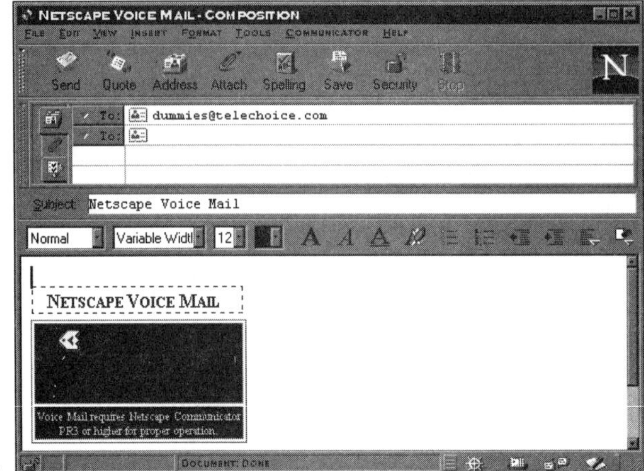

Figure 12-17: If you use Communicator for e-mail, you should recognize this screen.

6. **You can simply click on the Send button on the toolbar to forward the voice mail to the other party.**

   If you feel like changing subjects or adding other recipients, you can do so before clicking on the Send button. Just remember, if you add additional recipients, that they must be using Communicator for e-mail to be able to replay the voice mail message.

You don't have go through the calling process to send a voice mail message if you don't want to. Just choose Communicator➪Voice Mail from the menu bar. A dialog box pops up and asks you to enter an e-mail address. Fill in the text field and click on OK. The Netscape Voice Mail dialog box opens. Simply follow the previous steps (from Step 3 on) to send your voice mail message.

Getting your incoming messages

Incoming voice mail messages don't appear anywhere in Conference. Instead, they appear as new e-mail messages in Netscape Messenger. To find out whether you have voice mail waiting for you, just check your e-mail! The voice mail message appears as a new e-mail in your Messenger Inbox. When you open the message, a small Java applet runs and puts a playback control in the message. Just click on the arrow-shaped Play button to hear the message.

See Ya!

To quit Conference and exit the server, choose Call➪Exit from the menu bar. That's all, folks. . . .

Chapter 13

Picture This: Using CU-SeeMe 3.0

In This Chapter

- Take a peek at the warp coils
- Get set
- Go!
- Reflecting
- White(Pine)Boarding

CU-SeeMe 3.0 is White Pine Software's updated and enhanced version of the popular freeware CU-SeeMe program, which was developed at Cornell University (that's where the CU part of the name comes from). In fact, White Pine used the name Enhanced CU-SeeMe for previous versions of its product to show this distinction, though the name of this newest version reverts to just plain old CU-SeeMe.

The Windows 95 version of this product is still in beta testing as we write, but the final version will be available by the time you read this. White Pine plans to release a Mac version of CU-SeeMe 3.0 sometime later in 1997. So you Mac users (we feel your pain) must remain content just to read about the Windows version for now. The previous version of Enhanced CU-SeeMe (the 2.0 generation) is still available for the Mac, and it works just fine. However, the interface has completely changed with the advent of CU-SeeMe 3.0 — so although the general concepts remain the same, actually manipulating the program on your computer differs entirely from previous versions.

You can still find a freeware version of the program from Cornell; but if you take Internet video seriously, we expect that you're going to opt for the more functional, paid-for version because it isn't all that expensive and it has more capabilities. That's not to say that you can't try the Cornell version of CU-SeeMe — both programs can coexist happily on your hard drive, and the freeware version works very well — with the major limitation that it doesn't send or receive color video.

Much of what we discuss in this chapter does apply to the freeware version, however, so if you plan to download that program, go ahead and read this chapter. The interfaces of the two versions differ, and the Enhanced version has some different codecs (that make it more suitable for modem users), but in general, the two versions are similar. In case you're wondering, the different versions of CU-SeeMe — new and old versions of both White Pine's CU-SeeMe and the freeware Cornell version — can talk to each other, so you can use this new version to call or join a reflector conference with people who use the older versions.

A reflector site is a computer on the 'Net that is running special reflector software. This software allows the computer to act as a conference server — basically, the reflector receives the incoming video, audio, and chat data from each participant and sends it back out over the Internet to the other participants in the conference.

CU-SeeMe, which is not H.323 compliant, differs from all of the other programs we discuss in the previous chapters because it allows you to join into audio and video conferences with more than just one person at a time — not just a data conference, but actual audio and video on as many as 12 people at a time!

Note: Although CU-SeeMe is not H.323 compliant, it does include the H.263 video codec used by H.323 Internet telephony programs. White Pine is releasing a new generation of reflector software, called MeetingPlace, which allows users of NetMeeting and Intel Internet Video Phone to join into CU-SeeMe reflector conferences and share video with CU-SeeMe users.

The CU-SeeMe Interface

Before you get started, we want to take a minute to show you around CU-SeeMe. We understand that you probably don't enjoy going through all this interface and configuration stuff — and you're not alone. But this information is important, and you can't skip it — sort of like that foreign language requirement in school. So do read this section carefully because knowing (or not knowing) this information can make or break your Internet video experience. CU-SeeMe has two main windows:

- **The PhoneBook:** This window opens when you launch CU-SeeMe. This is the place where you can customize CU and from which you can make connections.
- **The Conference Room**: This window appears when you are in a call; it lets you see and control audio, video, and text chat.

When you start CU-SeeMe, the PhoneBook opens automatically. The Conference Room window opens only when you are actually in a call or on a reflector site.

The PhoneBook window looks like the one in Figure 13-1. It consists of a menu bar, a toolbar, a listing of PhoneBook entries (called Contact Cards), and a contact card display area.

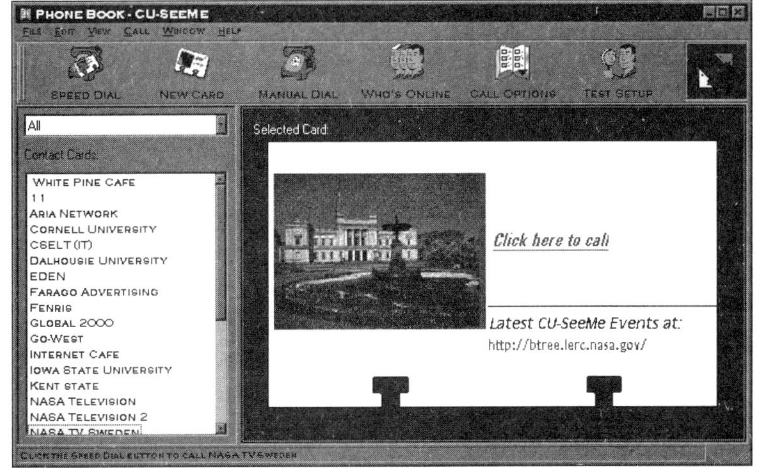

Figure 13-1: The PhoneBook.

The menu bar contains the following menus:

- **File:** Used to create Contact Cards and to exit CU-SeeMe
- **Edit:** A standard Windows Edit menu; used to open the various options dialog boxes
- **View:** Lets you customize what you see in the PhoneBook
- **Call:** Contains the controls for making calls to individuals and conferences
- **Window:** Lets you switch between the PhoneBook and the Conference Room (when it's available).
- **Help:** Provides access to the help system

Below the menu bar, a toolbar contains six icons (buttons). The toolbar has the following buttons (from left to right):

- **Speed Dial:** Enables you to quickly connect to a selected Contact Card
- **New Card:** Creates a new Contact Card
- **Manual Dial:** Lets you dial a person or conference that you don't yet have listed in the PhoneBook

- **Who's Online:** Provides you with a listing of all CU-SeeMe users who have logged into the Four11 CU-SeeMe directory service
- **Call Options:** Opens the Call Options dialog box, to let you set your preferences
- **Test Setup:** Lets you see (and hear) how your video and audio will appear when you make a connection

The Conference Room window, shown in Figure 13-2, is only available when you are actually engaged in a call with someone or have logged onto a conference. Across the top of the Conference Room, you find a menu bar containing the following menus:

- **File:** Lets you close the Conference Room window and exit CU-SeeMe
- **Edit:** Offers standard Windows editing tools, plus access to the Conference Room Customization dialog box
- **Conference:** Controls the display of video, audio, and chat and lets you hang up from the Conference
- **Window:** Enables you to switch back to the PhoneBook
- **Help:** Provides access to the help system

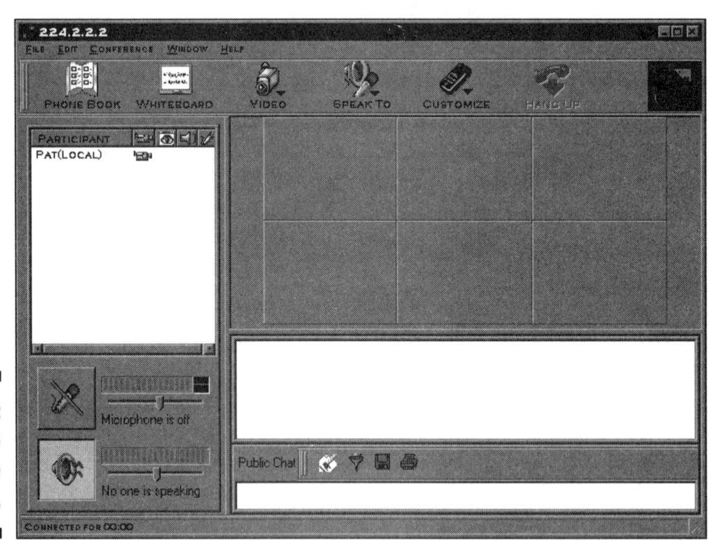

Figure 13-2: The Conference Room.

Directly below the menu bar, a toolbar offers the following buttons:

- **Phone Book:** Sends you back to the PhoneBook
- **Whiteboard:** Opens the WhitePineBoard
- **Video:** Provides a drop-down list of users in the conference who are transmitting video; select a name from the list to display the video
- **Speak To:** Lets you select a person in the conference to whom you wish to send audio. You can select an individual or send audio to all participants.
- **Customize:** Provides access to the Conference Room Customize dialog box
- **Hang Up:** Ends your current call

On the left-hand side of the Conference Room window, below the toolbar, you see the Participants List. This listing tells you who else is in the conference, and uses small icons next to each person's entry to show who has video, audio, and chat capabilities, as well as who is looking at your video.

You find the audio controls below the Participants List. By clicking on the large mike and speaker buttons, you can toggle your microphone and speaker on and off. The display bars provide a graphical representation of audio levels, and the sliders below the display bars allow you to control the audio levels.

To the right of the Participants List, you find the video display area — the place where you actually see the other people in the conference. If you've ever used another version of CU-SeeMe, this area will be quite a change for you, because the older versions display each person's video in a separate window. CU-SeeMe shows video in an array of video displays inside the Conference Room. The number of displays and their size depend on the preferences you select in the Video Layout tab of the Conference Room Customize dialog box. (We tell you how to work with this dialog box in a second.)

At the very bottom of the Conference Room window is the text chat area. This part of the Conference Room consists of a display area (the upper pane of the text chat region), a text entry field (the lower pane), and buttons that allow you to clear, filter (in other words, block), save, and print the text chat.

I've Got a Configuration, Jones

Quality may be Job One at Ford, but configuration is Job One in Internet videoconferencing. So we better get to it, right? Luckily, our job has been made easier by that familiar friend of the Internet telephony user — the Configuration Wizard. As long as you answer all of the questions that the Wizard asks you, you should be ready to start making calls right out of the box — with no additional configuration work. We do want to show you how to edit your profile and preferences, though — as you may wish to change things later on.

Setting things up

The Configuration Wizard purposely sets up things in a certain way — don't change anything unless you really need to. CU-SeeMe has lots of settings and options that you never need to touch — at least, we don't feel the need to do so during our use of the program. Of course, you probably don't want to change any of these settings until you gain more experience with the program. In the following discussion, we just talk about the settings that people typically need to change — and we skip over the ones that you shouldn't really be messing with unless you know what you are doing. (All you power users can read the manual if you feel the need to change a particular setting.) You typically deal with two configuration areas:

- **Your personal profile:** These settings define your identity on CU-SeeMe, as well as some general preferences, such as Internet connection speed.
- **Your calling options:** These options allow you to restrict access to CU-SeeMe by others (like your kids); you also can control how CU-SeeMe deals with incoming calls and how it displays your PhoneBook.

Editing your personal profile

Ready to get started? Okay — follow these steps:

1. **Choose Edit⇨Current Profile from the menu bar.**

   CU-SeeMe opens your current user profile — the one you set up when you first installed CU-SeeMe. If you want to set up additional user profiles, you can do so in the Call Options dialog box, which we discuss next.

2. **Click on the General tab.**

   Your screen should look something like the window shown in Figure 13-3.

3. **In the CU-SeeMe Name text field, enter the name you want other users to see when you are in a conference.**

4. **If you are connected to a network that can participate in Multicast conferences (probably not, for most dial-up ISP users), click on the Enable Multicast Connections check box to select it.**

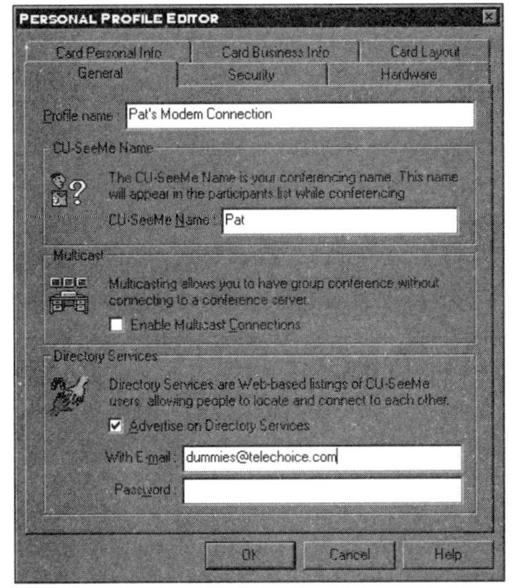

Figure 13-3: Name yourself!

If you registered with the Four11 directory service when you ran the Configuration Wizard, the Advertise on Directory Services check box should already be selected.

5. **Click on the Hardware tab.**

   You see a window similar to the one in Figure 13-4.

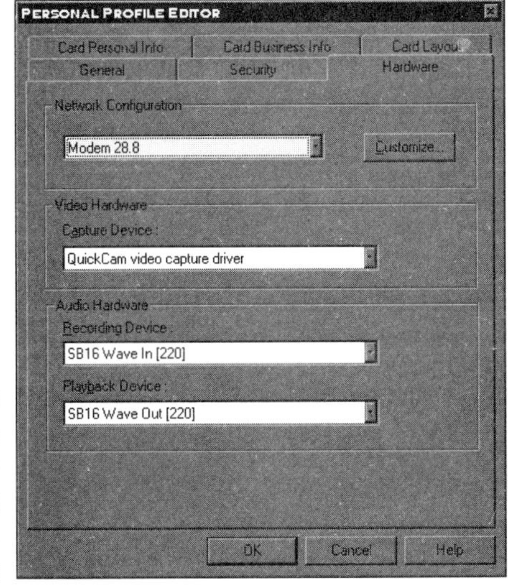

Figure 13-4: Setting your Network preferences.

6. **Make sure that the selection from the Network Configuration drop-down list reflects your Internet connection speed.**

   If your Internet connection speed doesn't correspond to one of the listed speeds, click on the Customize button and enter new maximum data receive and send rates in the dialog box that opens. Click on OK to close this dialog box.

7. **Click on the Card Personal Info or Card Business Info tab.**

   Which one you fill out depends on how you use CU-SeeMe — you can even fill out both if you'd like. We show you the Card Personal Info tab in Figure 13-5.

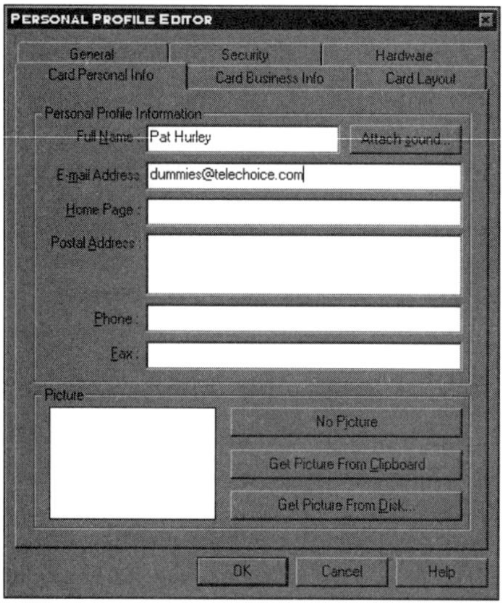

Figure 13-5: Enter information about yourself here.

8. **Fill out as much, or as little, information as you'd like.**

   Other people can download this information (your Contact Card) to their computers and add it to their PhoneBook, so if you want to keep some information private — like your phone number or e-mail address — leave these fields blank.

   You can customize the appearance of your Contact Card if you wish, by clicking on the Card Layout tab. Which Card Layout you choose is a matter of personal preference.

9. **Click on OK to close the Personal Profile Editor dialog box.**

Setting calling options

After you fine-tune your personal profile, you should verify your Call Options:

1. **Choose Edit⇨Call Options.**

   The CU-SeeMe Call Options dialog box opens, as shown in Figure 13-6.

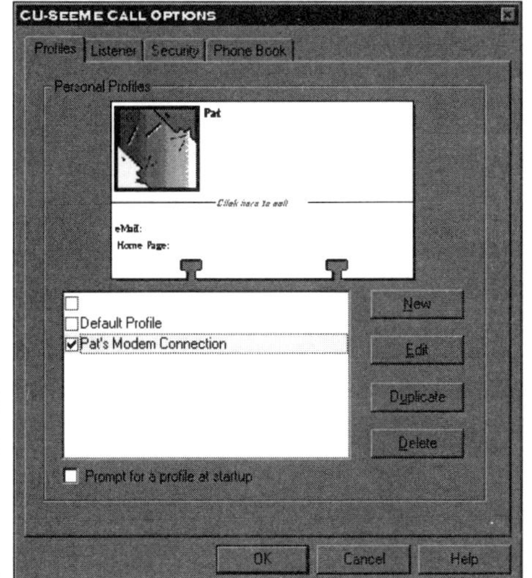

Figure 13-6: Setting your Call Options.

2. **Click on the Profiles tab.**

3. **The listing of Personal Profiles includes the profile that we show you how to edit in the preceding section of this chapter — make sure that a check mark appears next to this profile (a check mark identifies the profile you select).**

   You can create multiple profiles if you'd like — if your computer has more than one user, for example, or if you connect to the Internet in more than one way. (Perhaps you have a profile on your laptop for the times you connect via modem, and another profile — with higher network speeds — for the times when you connect via an Ethernet LAN connection at the office.)

4. **Click on the Listener tab.**

   Listener is a program that you can set up to run as soon as your PC starts up. It allows your computer to monitor your network connection for incoming CU-SeeMe calls even if you don't have CU-SeeMe running.

Control panel fun

You can configure and customize CU-SeeMe in almost limitless ways. As evidence of this flexibility, we present you with the CU-SeeMe Control Panel. Found in the Windows Control Panels folder (choose Settings⇨Control Panels from your Windows Start menu to find this folder), this Control Panel lets you set up your Profile, Listener, and Security preferences without opening CU-SeeMe. Why? Well, we're not exactly sure ourselves — but it does make sense if you want to change these things while you're not connected to the Internet. You can either find this to be an interesting and useful tidbit or file it away in the "not so useful information" section of your brain. Your choice.

5. **If you want Listener to check for incoming calls from the time you start your computer, click on the Run listener at startup check box to select it.**

   You may find yourself receiving incoming calls you don't want from overeager CU'ers — so think twice about running Listener.

6. **Click on the Security Tab and select the Parental Control options you want (if any).**

   These controls, which are password protected, let you restrict access to CU-SeeMe. For example, you can completely block access to CU-SeeMe, or you can allow your children limited use — so they can call the reflectors and people listed in your PhoneBook, but not make direct calls or receive incoming calls.

7. **Click on the PhoneBook tab.**

   This tab lets you customize what is shown in the PhoneBook. We usually keep everything visible, but you can pare the PhoneBook down according to your own personal preferences.

8. **Click on OK to save your changes and close the CU-SeeMe Call Options dialog box.**

That wasn't too painful was it? You can always go back and change any of these preferences at any time, but you should be ready to go ahead and start CU'ing!

Testing it out

Before you start making calls, you should run through the built-in Test Setup system to make sure that your video and audio work properly. To run the Test Setup system, simply follow these steps:

1. **In the PhoneBook, click on the Test Setup button on the toolbar.**

   The Test Setup dialog box opens, as shown in Figure 13-7.

   On the Test Video tab, you can see your own video, both local (what your video device is feeding into CU-SeeMe) and remote (what someone on the other end of a call would see).

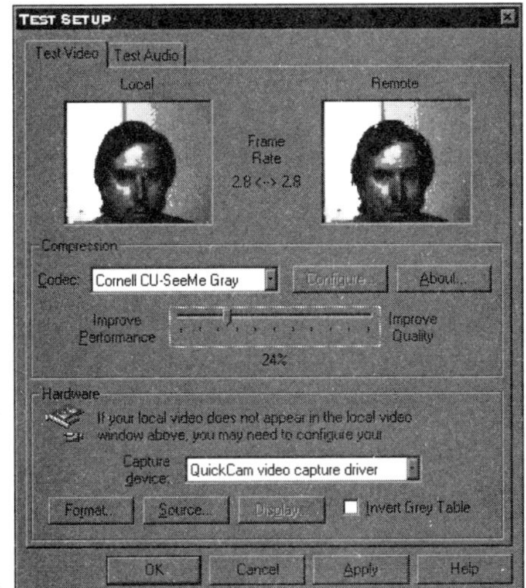

Figure 13-7: See what you'll look like online.

2. **From the Codec drop-down list, you can select a different video codec to try out.**

   If you use a modem-speed Internet connection, you should really stick with the White Pine Color codec — or even the Cornell CU-SeeMe Gray codec if you plan to communicate with people who use the freeware version of CU-SeeMe and you want them to be able to see you. The other codecs (such as White Pine H.263 and White Pine M-JPEG) are higher-quality codecs that are optimized over high-speed connections, like Ethernet LANs.

   Although this may seem counterintuitive, the White Pine Color codec is actually so much more efficient than the Cornell CU-SeeMe Gray codec that it uses less bandwidth — despite sending full-color video. So select the White Pine Color codec as your default codec unless you have a reason to go with the grayscale codec.

3. **Click and drag to adjust the Compression quality slider.**

   If you use a modem connection, you should be somewhere on the left-hand side of this scale — say around 15–20 percent. This setting allows you to maintain a reasonably high frame rate for your video. Higher-quality settings may make your picture look a little bit better, but you sacrifice movement and your video will look more like a still image.

4. **If you can't see your video, or if you have more than one video capture device, and want to switch to a different one, you can select it in the Hardware section of this dialog box, by using the Capture device drop-down list.**
5. **Click on the Test Audio tab.**

   Figure 13-8 shows the Test Audio tab.

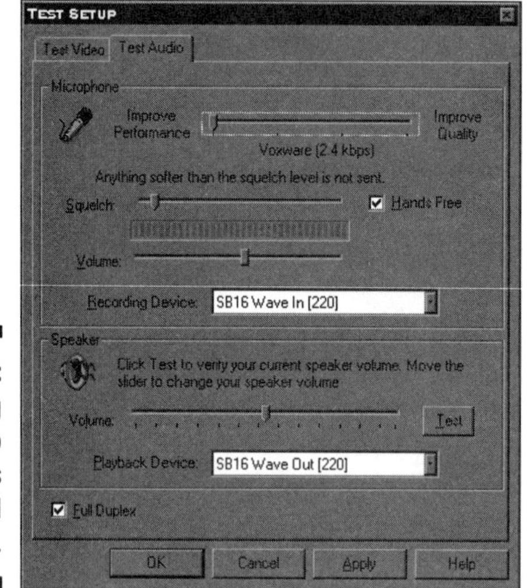

Figure 13-8: Getting your audio settings straightened out.

6. **In the Microphone section of the dialog box, use your mouse to move the slider from left to right.**

   Moving the slider changes your selected audio codec — the codec currently selected by the slider is listed underneath the slider. (For example, the Voxware codec is selected in Figure 13-8.) For modem connections to the 'Net, you want to use one of the two lower bandwidth codecs (the Voxware codec or the DigiTalk codec), which you select by keeping the slider toward the Improve Performance side of the scale.

   We like to leave as much bandwidth as possible for video, so we usually choose the Voxware codec.

7. **Click the Hands Free check box to let CU-SeeMe determine when to activate your microphone — otherwise, you need to manually click on the microphone button to send audio during a conference.**
8. **Speak normally into your microphone and make sure the microphone display level moves into the green bars when you speak.**

If it doesn't, slide the Squelch slider to the left until the green bars light up while you talk. Don't move the slider so far to the left that the green bars are activated by background noise in your room — if this happens, you'll be sending audio all the time and making people on the reflector sites you visit very unhappy.

9. **In the Speakers section of the dialog box, click on the Test button to set your audio speakers or headphones volume level.**

   If the test message you hear is too loud, move the Volume slider to the left. If it's too soft — nah, we won't tell you, you can figure that out yourself.

10. **If you have a full-duplex sound card, make sure the Full Duplex check box is selected.**
11. **Click on OK to save your changes and close the Test Setup dialog box.**

Videoconferencing for (And with) Everyone

CU-SeeMe lets you videoconference in one of three ways:

- Person to person
- With several people on a reflector site
- With several people via a multicast conference

For most users (we mean you!), the first two methods are by far the most common. Multicast conferences (which use the same sort of Internet transmission scheme as the multicasting we talk about in Chapter 8 — in the discussion of streaming video products) typically aren't available to the average Internet users, because most ISPs and other Internet access providers aren't configured to support multicasting. In all likelihood, this method of conferencing can be used by people on a corporate intranet that has the latest and greatest hardware to support multicasting, or used by the lucky few Internet users who are connected to a multicast-capable section of the Internet.

Regardless of what type of conference you want to join, you have two basic ways of making the connection:

- Using a Contact Card
- Dialing directly to an IP address or host name

You can also search for, and call, individuals who have listed themselves on the Four11 directory, using the Who's Online function.

Making a call from the PhoneBook

The best way to get started with CU-SeeMe and to get a feel for the CU-SeeMe community is to log onto a CU-SeeMe reflector site. As we discuss in Chapter 6, a *reflector site* is basically a server on the 'Net that someone has equipped with CU-SeeMe reflector software. This software comes in two flavors — the freeware CU-SeeMe reflector and the Enhanced CU-SeeMe reflector — and you can log onto servers that use either version, although the freeware reflectors do not support color video. If you do log onto a freeware reflector using one of the color video codecs, people can hear you but not see you — so switch to CU-SeeMe Gray in the Call Options dialog box.

After you log onto the reflector site, you can see and be seen by everyone on the site. Well, that's not completely true — actually, everyone on the site who wants to see you can do so (just by opening your video window), and you see everyone who is transmitting video, up to the limit of open video windows that you specified while setting your preferences.

White Pine has set up a reflector just for you, the first-time user. Log onto this one just to get acquainted with how everything works. After that, you should be ready to dig up your own list of favorite sites and start exploring.

Here's how you log onto the White Pine reflector site:

1. **Establish a connection with your ISP.**
2. **Start CU-SeeMe by double-clicking on its icon or by choosing CU-SeeMe from the Start menu.**
3. **In the PhoneBook window, scroll through the listing of Contact Cards until you find White Pine (Public), and click on it.**

   The White Pine (public) Contact Card is displayed in your PhoneBook, as shown in Figure 13-9.
4. **Click on the "Click here to call" section of the Calling Card or click on the SpeedDial button on the toolbar.**

   A Connection dialog box opens on your screen, as shown in Figure 13-10.
5. **Enter** 3 **in the Conference ID field or click on Conference 3 in the listing that's displayed.**

   You don't have to try just Conference 3 — as you can see, White Pine has other available Conferences. Some reflectors have only one conference, in which case you don't have to decide which one to enter.

   Leave the Password text box blank. Some (private) reflectors use this text box to restrict entry.

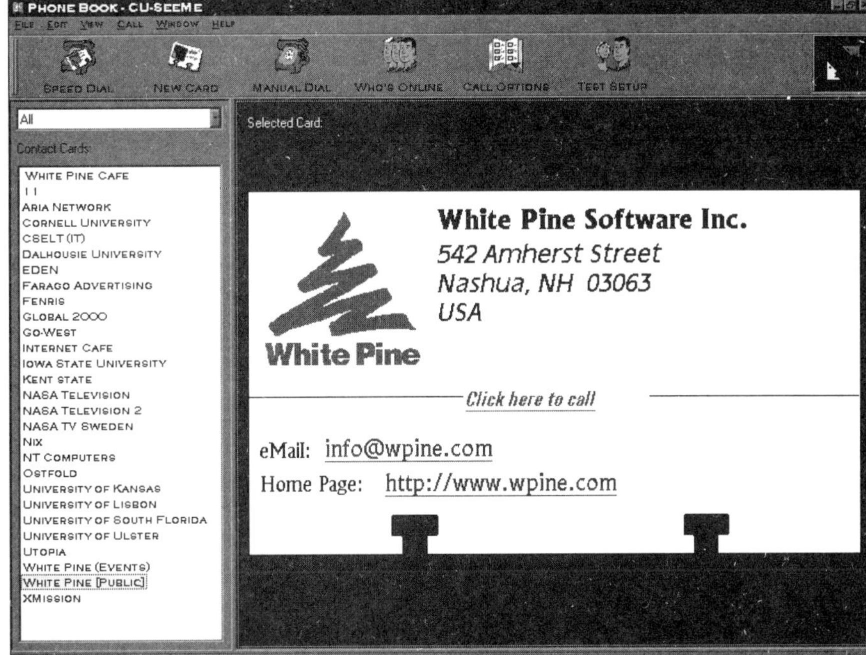

Figure 13-9: White Pine's Reflector Contact Card.

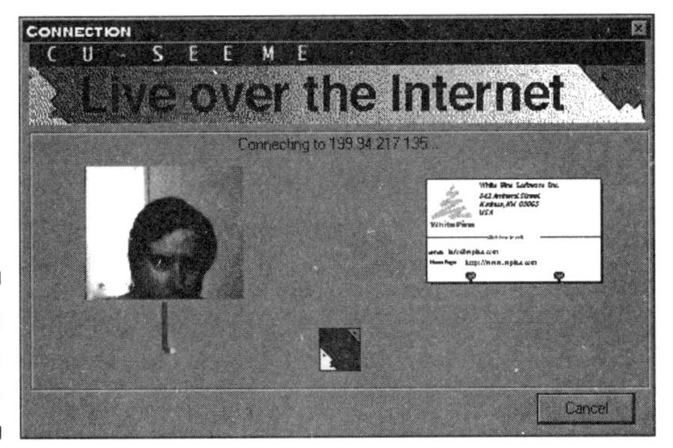

Figure 13-10: Making the connection.

6. **Click on the Join button.**

   This action should connect you to the reflector. Many of the public reflectors can be quite busy at times, so don't be surprised if you don't get in on the first try.

After you connect, you're greeted by the Message of the Day. This message typically tells you a little about the reflector and may even lay down some guidelines, such as no nudity. (Yes, that can be a site on some reflectors, but not on public ones, for the most part — and no, we're not going to tell you where to look.) The message may also note the maximum bandwidth that you can send (not a problem for modem users).

If you see this Message of the Day, click on the Go button to enter the reflector. Not all reflector sites display a Message of the Day, in which case you automatically enter into the conference.

7. **The Conference Room window opens, as shown in Figure 13-11, and you're connected!**

You use this same process to call an individual person for whom you have a Contact Card, except that you don't see a dialog box asking you for a Conference ID or Password — if the other person accepts your call, you move right into the Conference Room.

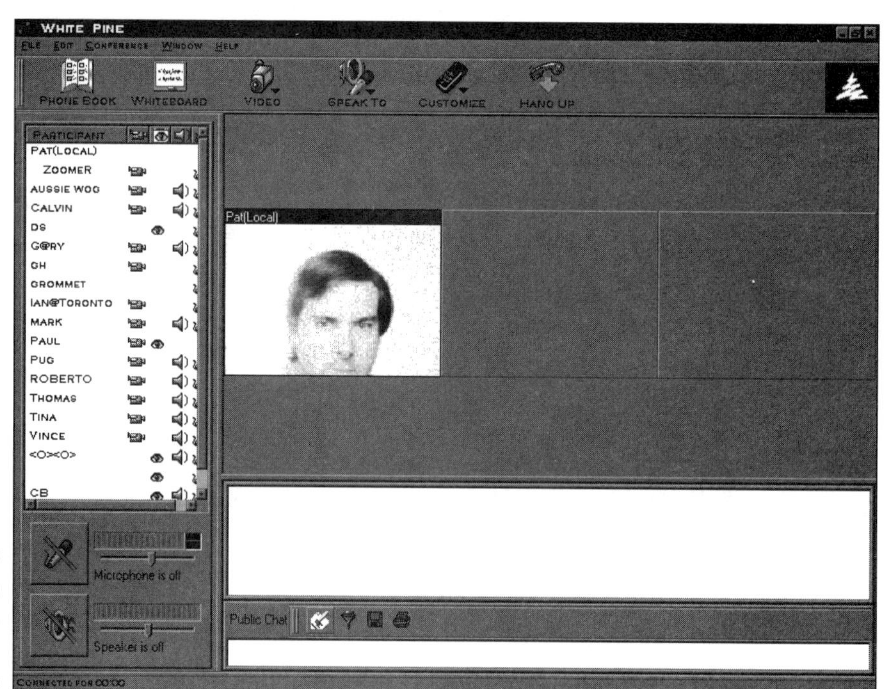

Figure 13-11: You're in the conference!

Looking, talking, making adjustments

After you're connected, you can navigate around the conference and find out who you want to see or talk to and who you don't.

Making the Conference Room your own

Like everything else in CU-SeeMe, the Conference Room can be customized. In fact, you have a plethora of options for making the Conference Room appear, and behave, exactly as you want. To access these controls, simply click on the Customize button in the toolbar and select one of the options that pop up. It doesn't really matter which one you select at first — because all of the Customize controls are located on tabs in the same dialog box — but if you know exactly what you want to change, you can go right to that tab. Figure 13-12 shows the Customize dialog box. We don't go through a detailed discussion of what you can do with each of these tabs, but rather give you a brief rundown on each of them:

- **General:** This tab lets you specify the general layout of the Conference Room.
- **Video Layout:** You'll probably visit this Customize tab most often, because it lets you control how many videos CU-SeeMe can show on your screen at one time. To increase the number of video windows, drag the slider to the right with your mouse — to decrease, drag to the left.

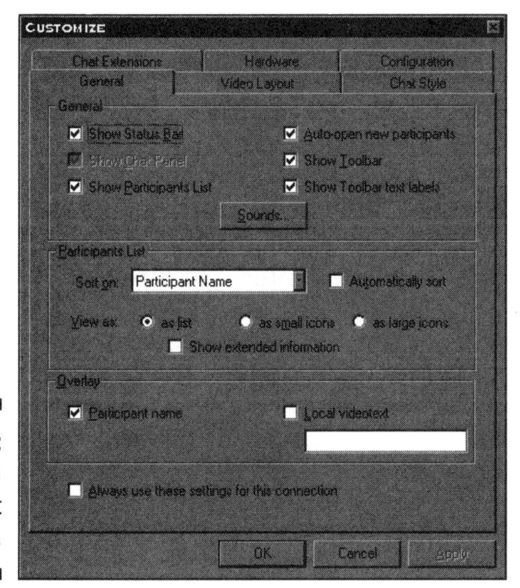

Figure 13-12: Customizing the Chat Room.

Just watching . . .

You can use CU-SeeMe as a broadcast viewer or as a way to join reflector conferences without sending any video (that is, lurking). Some of the more popular reflector sites are, in fact, actually broadcasters — NASA Select (watch the space shuttle and Johnson Space Center feeds being transmitted from several sites) and ABC World News Now (WNN), for example, are transmitted via CU-SeeMe reflectors.

White Pine is developing super-big and fast server technologies that should enable CU-SeeMe to be used as a 'Net broadcast medium, competing with the streaming video technologies that we discuss in Chapter 8.

Logging onto a broadcast reflector as a lurker is like logging onto any reflector as a participant. The only difference is that you must uncheck the Send Video check box in the Contact Card editor or when you use the Manual Dial dialog box. Make sure that you do so before you log on — some reflectors don't accept connections if you're trying to transmit.

Note: As long as you have a camera, you can switch back and forth between lurker and participant status on the fly if you're connected to a reflector. Sometimes, however, the group that you're not in (whether as lurker or participant) is already full, which you can't tell unless you know the limits ahead of time. But if you try to switch to the other group and it's full, you're disconnected. Don't say we didn't warn you.

Although you may want to have as many videos on your screen as possible, remember that the more you show, the less bandwidth each video stream has — which means that you end up with lower quality and lower frames-per-second rates for everyone. You also end up with smaller video frames and put more of a strain on your CPU as it tries to keep up with decoding all of that incoming video. Modem users are usually better off if they keep the maximum number of video frames to about four or five.

- **Chat Style:** You can change the font and font size of text in the Chat window, select keywords to be highlighted if they appear in the chat, and allow URLs to be highlighted and launched in your preferred application if you wish. For example, if you choose to allow URL launching and someone types a Web page URL into the text chat, you can double-click on that URL and launch your Web browser to view the page. Pretty neat.
- **Chat Extensions:** This tab lets you create macros so that you can use keyboard shortcuts to type things you often put into text chat — for example, the URL of your home page or your e-mail address.
- **Hardware:** This tab lets you adjust your Network settings (that is, your maximum send and receive data rates) and lets you control your audio and video hardware settings.

- **Configuration:** You can go into this tab to reset or change your audio and video codec settings on the fly — in case you need to switch to grayscale, for example, or if you need to move to a lower bandwidth audio codec to improve audio transmission.

Checking out the Participants List

Probably the most useful and important part of the Conference Room is the Participants List. From this list, you can see who's in the reflector, who's sending video, who's looking at your video, and who can send or receive audio. That's just the tip of the iceberg, actually — you can use the Participants List to find out a lot more about your conference partners, including what version of CU-SeeMe they use, data statistics, and IP addresses.

Note: We discuss the Participants List as you see it when viewing it as a list. You can also display the Participants as icons — which changes the way CU-SeeMe displays this information. You can switch your view of the Participants List in the General tab of the Customize dialog box — we generally prefer the list view because it gives you the most information at a glance.

The Participants List contains the following information:

- Each person's CU-SeeMe Nickname.
- Icons to indicate which persons are sending video.
- Icons to identify which persons can receive your video. An open "eye" inside the icon identifies a person who has your video window open on his or her screen.
- Speaker icons to identify those parties who can send audio.
- A microphone icon to identify a person to whom you can send audio.

Displaying video

When you first enter the reflector, you see some video windows pop open. The number of windows that open are the lesser of the following two figures:

- The number of people connected to the reflector and sending video
- The maximum number of video windows that you selected during your setup

If you don't want to have videos open automatically like this, open the General tab in the Customize dialog box and deselect the Auto-open new participants check box by clicking on it.

To open someone's video manually, click on the Video button on the toolbar and select that person's name from the list that pops up. If you want to stop viewing a person's video, just repeat this process to deselect the person's name from the Video button's pop-up menu.

You probably want to see yourself as well — your "local video." To do so, simply select yourself from the Video button's pop-up menu.

Tuning in to audio

Although the concept may seem counterintuitive, we often find that decent audio is actually much more important than video quality during a videoconference. If the audio isn't of good quality, people often opt for using the Chat window or calling on a separate telephone line by using the PSTN. You often find only a few diehards using audio, whereas the rest of the people in the conference use the Chat function. And audio during a multi-party conference can get a little confusing, especially if a whole bunch of people are talking at once.

To send audio, just speak into your microphone. That's all.

If you don't select hands-free audio when you run your Self Test (as we describe in a previous section of this chapter), you have to click on the microphone button before you talk.

If you want to send audio to only one individual, instead of the whole reflector, follow these steps:

1. **Click on the Speak To button in the toolbar and select the name of the person with whom you wish to speak privately.**
2. **Talk into your microphone.**
3. **When you are done speaking privately, click on the Speak To button again and select "everybody."**

If you use the lowest bandwidth codec (Voxware) and you still can't get decent audio quality, try freezing your video before you speak. Right-click on your Local Video display and choose Freeze Video from the pop-up menu that appears. You then can devote much more bandwidth temporarily to your audio signal. Repeat this process to unfreeze your video.

Chat me up

If the audio is just too much of a zoo, or if you're on a bad 'Net connection and you just can't make it work, you can use the Text Chat feature to communicate.

To enter something in the Chat window, type your message into the bottom pane of the window and press the Enter key on your keyboard. Notice that your name accompanies whatever you write, as other people's names accompany their own text. This feature helps you understand who is saying what.

You can send chat to the entire reflector as well as to individuals — which can be handy if you want to share some sort of private information (like an e-mail address or phone number) with a friend on the reflector, but don't want to broadcast it to everyone. To send chat privately:

1. **Choose Conference⇨Chat To⇨ and the name of the person with whom you wish to have a private chat.**

   Both parties must complete this step to have a two-way private chat. If you send private chat to someone and that person doesn't select you for private chat, CU-SeeMe sends that person's replies to everyone on the reflector, and not just to you.

   A check mark appears next to the name of the person you select, to indicate that you are in a private chat session.

2. **Type your chat messages normally in the bottom pane of the Chat window.**

   Only the person you selected receives your chat.

3. **When you complete your private chat session, repeat Step 1 to deselect the other person.**

In some older, freeware versions of CU-SeeMe, the Chat tool was called CU-SeeMe Talk. So if you hear people mention Talk, they may not mean audio.

Hanging up

The time comes in every good reflector session when you need to say good-bye and go back to your real life, although some folks just can't seem to get enough CU-SeeMe. (Or *any* of these programs, for that matter — after a while, you notice a few names that seem to never leave the active Participants Lists.) So unless you plan on sleeping in front of the camera, you need to disconnect. To do so, just click on the Hang Up icon on the toolbar or choose Conference⇨Hangup from the menu bar. If you're really slick, you can press Ctrl+K on your keyboard.

Be direct!

You don't need to use a Contact Card if you want to have a conference with someone or join a reflector — you can call that person or reflector directly by IP address or host name. To place a direct call from the PhoneBook, follow these steps:

1. **Choose Call⇨Manual Dial from the menu bar or click on the Manual Dial button on the toolbar to open the Manual Dial dialog box, as shown in Figure 13-13.**

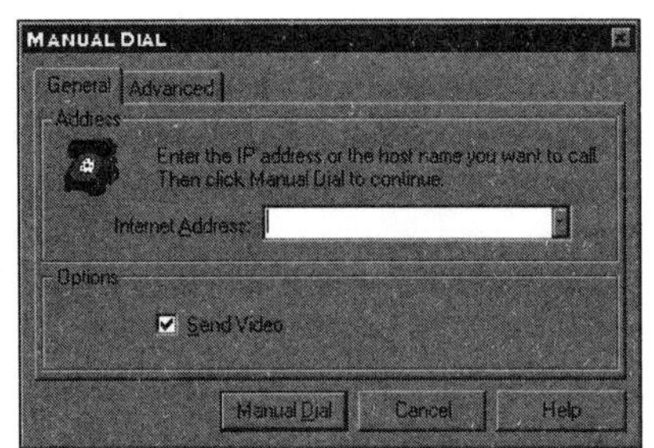

Figure 13-13: Dialing an IP address.

2. **In the Internet Address text field on the General tab of the Manual Dial dialog box, type the IP address or host name of the person or reflector you want to call.**
3. **If you wish to send video, select the Send Video check box.**
4. **If you are calling a reflector that has a Conference ID and password, click on the Advanced tab and enter these values in the Conference ID and Password text fields.**

   This step is optional.
5. **Click on the Manual Dial button to place the call.**

That's all you need to do. The rest of your call proceeds exactly as the Contact Card call we discuss earlier in this chapter.

Checking out who's online

CU-SeeMe includes a program which, if you choose to use it, automatically logs you into the Four11 CU-SeeMe directory service. This is a handy way to find a friend or acquaintance if you don't know his or her IP address and you don't want to search through your favorite reflectors.

To place a call through the Four11 directory, just follow these steps:

1. **From the PhoneBook, open the Who's Online dialog box (see Figure 13-14) by clicking on the Who's Online button on the toolbar.**
2. **Scroll through the listing until you find the person you want to call, and select that person's name by clicking on it.**

   You can sort the listing in the dialog box by clicking on one of the column headers (Name, E-Mail Address, and IP Address).

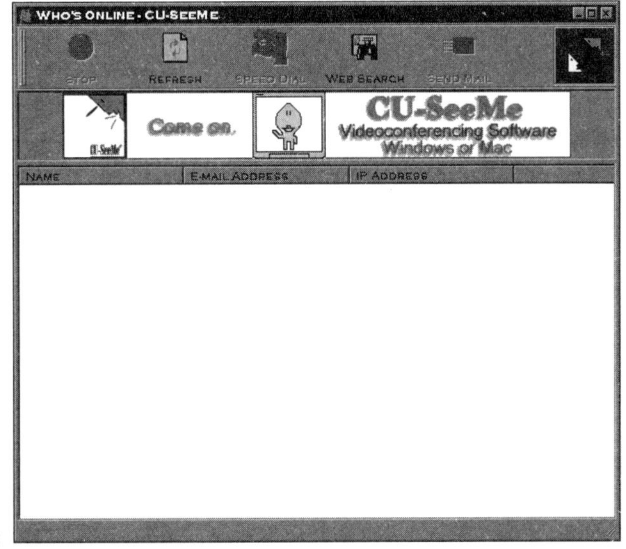

Figure 13-14: Seeing who's online.

3. **Click on the Speed Dial button in the dialog box toolbar to initiate the call.**

   The Connection dialog box opens, just as it does in other calls, and CU-SeeMe connects your call.

Keeping Track of Addresses

Make sure you get very familiar with the CU-SeeMe PhoneBook. After all, it's the main interface to the program — and it's where many of your calls originate. You don't have to just stick with the reflectors and people that come prelisted in the PhoneBook, however. You can easily add new Contact Cards to customize your PhoneBook.

You can add new Contact Cards to the PhoneBook in two ways:

- By using the Contact Card Assistant
- By retrieving a person's Contact Card while you are connected to that person in an individual conference or on a reflector

The second option works only if the other person is also using CU-SeeMe 3.0 — earlier versions of CU-SeeMe do not use Contact Cards.

To create a new Contact Card using the Assistant:

1. **From within the PhoneBook, click on the New Card button on the toolbar or choose File⇨New Contact Card Assistant from the menu bar.**

   The Contact Card Assistant dialog box opens.

2. **Click the Next button to begin.**

   The Enter Calling Information screen opens in the dialog box, as shown in Figure 13-15.

3. **Fill in the blank text fields.**

   You must add at least the e-mail address or IP address of your contact. Typically, you would add the e-mail address of a person who does not have a fixed IP address, so you can use the Four11 directory to locate his or her IP address each time you call. Reflectors, on the other hand, have fixed IP addresses, so you would enter this address instead of an e-mail address.

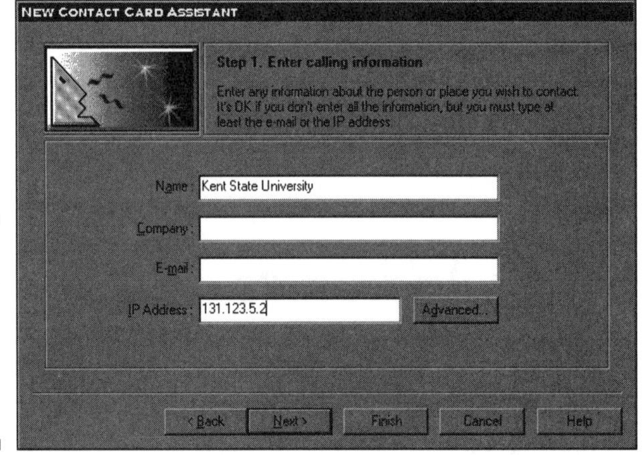

Figure 13-15: This is the vital information for a Calling Card.

   If you know that the reflector has a particular Conference ID or Password, you can enter this information by clicking on the Advanced button next to the IP address text field. A Conference Server Information dialog box opens, to allow you to enter these values. If you don't enter these items now and they are required by the reflector, you will be prompted for them when you try to connect.

   You are actually finished now — at least with the minimum requirements. The information you have added is all you really need to use the Contact Card to make connections. If you don't want to mess around with additional data or pictures, just click on the Finish button. If you want to enter that additional stuff, keep following the instructions.

4. **Click on the Next button twice (you do this a lot throughout the process).**
5. **Click on either the Personal information or the Professional information radio button, depending on whether this is a personal or professional contact.**
6. **Click on the Next button again.**

   The Enter data screen of the dialog box opens, as shown in Figure 13-16.

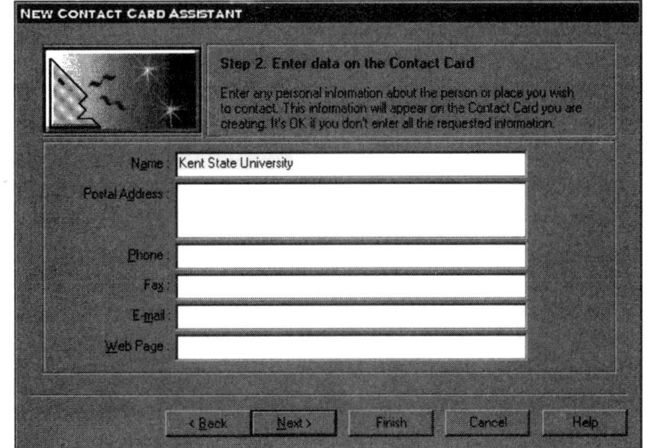

Figure 13-16: These details are all optional.

7. **Enter whatever information you wish to put in here (it's all optional) and click on the Next button two more times.**

   You find yourself in the Add a picture screen of the dialog box.
8. **You can either browse your hard drive for a picture (click on the Find a picture radio button and then on the Browse button) or paste one from the Windows clipboard (select the Use a picture on the Clipboard radio button).**
9. **Click on the Finish button, and you're done.**

If you are in a person-to-person or reflector conference with someone who uses CU-SeeMe 3.0, you can retrieve Contact Cards without having to enter anything at all. Here's all you have to do:

1. **Choose Conference⇨Get Contact Card Of⇨ and the person's name.**

   The Downloading Contact Card dialog box opens, to show the status of the Contract Card retrieval.

2. **When the dialog box indicates that the Contact Card has been downloaded, click on the Save button.**

   The Save Contact Card As dialog box opens. You can change the Nickname of the Card if you wish, by typing a new entry in the Nickname text field.

3. **Click on OK to save the Contact Card to your PhoneBook.**

Contact Cards that you download are only as good as the information that the other person puts into them. If the other person doesn't fill out the profile completely or correctly, the Contact Card that you download may not be particularly useful.

After you've made PhoneBook entries, you can always go back and change the information in the Contact Card. For example, if you've added a new reflector to your PhoneBook and you find out that it's a broadcast reflector — where you shouldn't be sending any video — you can edit the Contact Card to reflect (no pun intended — really) this change. Other reasons to edit a contact card include:

- **Changes in network settings:** The Contact Card Assistant sets up new Cards using your default network settings from your user profile, but some reflectors have specific limits on the maximum or minimum Send and Receive rates.
- **Change in IP address**: It's not unheard of for a reflector to move to a new IP address — something you can easily change in the Contact Card Editor.
- **Changes in Conference ID or Password:**, To retain their reflector's privacy and exclusivity, reflector operators often change the conference ID and password.

Editing a Contact Card is easy. Starting in the PhoneBook window:

1. **Right-click on the Contact Card's listing and choose Properties from the resulting pop-up menu.**

   The Contact Card Editor dialog box opens.

   You can change any of the previously entered information in the Contact Card in this dialog box. For example, to change the contact information, you can click on the Card Personal Info or Card Business Info tabs. We're going to concentrate on the Connection and Network tabs — these are the ones that most often need fixing or adjusting.

2. **Click on the Connection tab.**

   The Connection tab moves to the top of the dialog box, as shown in Figure 13-17.

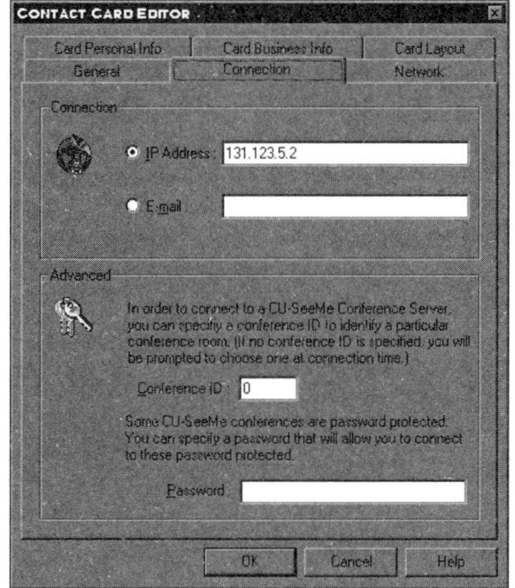

Figure 13-17: The connection details.

3. **Enter any updated or corrected IP addresses, Conference IDs, and Passwords into the respective text fields.**
4. **Click on the Network tab to bring Network options to the front, as shown in Figure 13-18.**

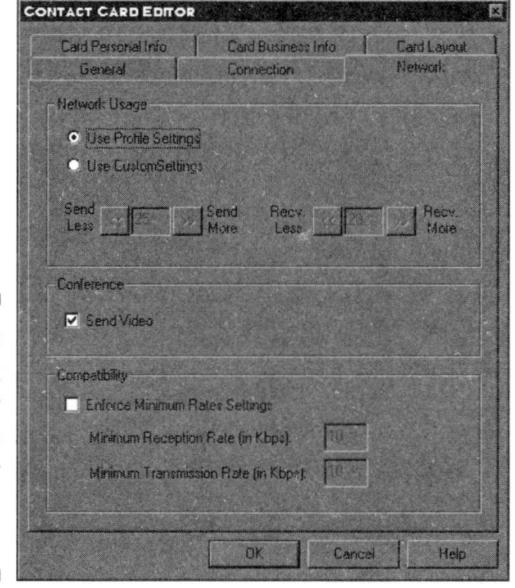

Figure 13-18: Network options for this particular Calling Card.

5. **If you need to raise or lower your Send and Receive rates, click on the Use Custom Settings radio button.**

6. **Change your maximum Send and Receive rates by clicking the up or down arrows next to each setting.**

7. **If the reflector site you are editing does not accept incoming video (for example, if it is a NASA TV site), click on the Send Video check box to clear it.**

   Occasionally, but rarely, reflector sites require you to lower your minimum transmission and reception settings. You can do this in the Compatibility section of the dialog box by simply entering these lower numbers in the appropriate text fields.

8. **Click on OK to save your changes and close the dialog box.**

You've only changed the settings for this particular Contact Card — the default settings for your user profile have not changed. These changes take effect only when you initiate a call to this particular person or reflector site.

CU-SeeMe and the Web

Like many of the products we discuss in this book, CU-SeeMe can work in cooperation with your Web browser. The degree of this integration varies from product to product — in the case of CU-SeeMe, reflector operators can put links to their sites on a Web page, and users can find the site on the page and click on the link. Then CU-SeeMe automatically launches and connects to the reflector site.

You can also launch CU-SeeMe from a Web site. After you run across a link to a CU-SeeMe reflector on a Web page, you can just click on the link to join that conference. All the details about the connection (the IP address, conference number, and so on) are transmitted to CU-SeeMe automatically. Before you can use this easy process, however, you need to go into your browser's Preferences menu and add CU-SeeMe as a helper application to make this connection work. Here's what you need to do (using Netscape as an example — other browsers function similarly):

1. **Start Netscape.**

2. **Choose Options⇨General Preferences from the menu bar.**

   Netscape displays the Preferences dialog box.

3. **Click on the Helpers tab in the Preferences dialog box.**

4. **Click on the Create New Type button.**

   The Edit Type dialog box opens.

5. **In the File/MIME Type text box, enter application.**

6. **In the Subtype box, enter x-cu-seeme and then click on OK.**
7. **In the File Extensions box, enter cu,csm.**
8. **Click on the Launch the Application radio button, to select it.**
9. **Click on the Browse button, use the resulting browse dialog box to find CU-SeeMe on your hard drive, and after you find CU-SeeMe, double-click on it.**

   When you're all done, you see something like what's shown in Figure 13-19.

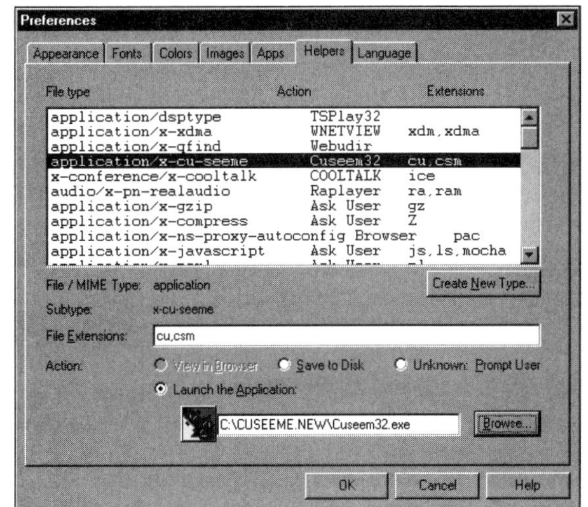

Figure 13-19: Netscape is now set to launch CU-SeeMe.

10. **Click on OK to save your changes and close the Preferences dialog box.**

CU-SeeMe can use your Web browser to find — and place calls to — people who are listed with the Four11 CU-SeeMe directory service. To launch your browser and go to the CU-SeeMe directory search page, simply choose View⇨Directory Services from your CU-SeeMe PhoneBook menu bar. Your preferred Web browser launches and automatically opens the search page on Four11's Web site.

The WhitePineBoard

The *WhitePineBoard* is a version of a multimedia collaborative application (in fact, it's actually an implementation of a program called FarSite by DataBeam Corporation) that enables people involved in a CU-SeeMe conference to share information while conferencing. The WhitePineBoard enables you to perform the following tasks:

- Work collaboratively across the 'Net
- Import and display documents from most word-processing, spreadsheet, and graphics applications
- Transfer files
- Use drawing tools and text to mark and annotate the image that you're viewing
- Capture portions of the screen or desktop for display
- Use the WhitePineBoard to share applications, like you can do in Microsoft NetMeeting

The WhitePineBoard actually runs as a separate program — you can even run it without using CU-SeeMe if you'd like. If you've downloaded CU-SeeMe from the White Pine Software Web site, you may not have even downloaded the WhitePineBoard yet — it's a separate (and rather large) download. To begin a WhitePineBoard conference, follow these steps:

1. **Connect via CU-SeeMe to the other party or to the reflector for a multiparty conference.**
2. **Have all parties open the WhitePineBoard (see Figure 13-20) by clicking on the Whiteboard icon on the toolbar or by choosing Window⇨Whiteboard from the menu bar.**
3. **After you've finished using the WhitePineBoard, you can close it by choosing File⇨Exit from the WhitePineBoard menu bar.**

After you finish using CU-SeeMe, all you need to do is quit the program. You already left the reflector or your person-to-person call by clicking on the Hang Up icon. So just choose File⇨Exit from the menu bar, and you're done.

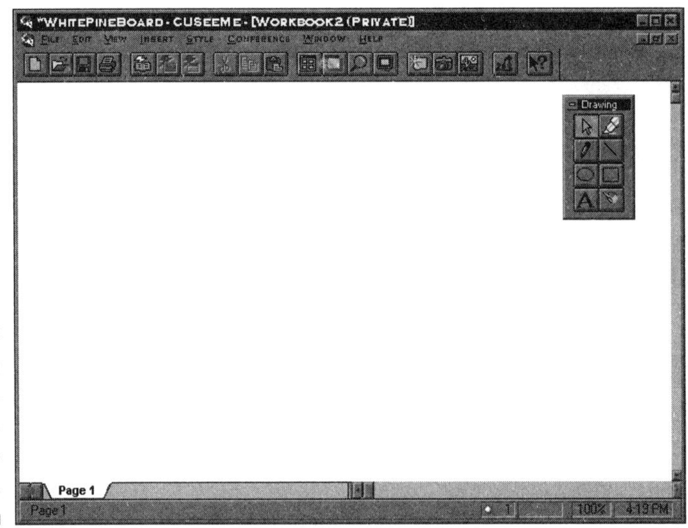

Figure 13-20: Collaborate with the WhitePineBoard.

Part III

Internet Telephony: Not Just for PCs Anymore

" WELL, I'M REALLY LOOKING FORWARD TO SEEING THIS WIRELESS E-MAIL SYSTEM OF YOURS, MUDNICK."

In this part . . .

Internet telephony is rapidly moving beyond the limitations of the PC-to-PC software that made up the first generation of products. Now you make calls to regular phones from your PC, send faxes across your corporate intranet, and even call from telephone to telephone via the Internet without ever touching a PC.

In this part of the book, we get you up to speed on the services that offer Internet calling to regular telephones, and the gateway hardware you can add to your corporate network to make calls and send faxes between offices without running up a huge long-distance bill. We also talk about some of the new services being rolled out by telecommunications companies — big companies with names you know — that are making the distinctions between Internet telephony and regular telephony increasingly hard to discern.

Chapter 14

PC-to-Telephone Gateway Services

In This Chapter

- Call Grandma from your PC
- You gotta pay to play (well, to talk)
- Where to find PC-to-phone gateways
- Do you want the XXL in stone or fuscia?

In the previous chapters of this book, we focus on using Internet telephony to talk from one PC to another across the 'Net. Until recently, this was the main — almost the only — way of talking via the Internet. A big change occurred late last year, though, when first VocalTec (the company that makes Internet Phone) and then a host of other companies began to market Internet telephony gateways.

As we discuss in Chapter 2, gateways are basically magical black boxes that connect on one end to the Internet and on the other to the regular PSTN (Public Switched Telephone Network). The black box is magical because it contains computer hardware and software (which is, by definition, magical) that convert the packets of digital 'Net telephony audio into the analog signals used by the PSTN.

Gateways can be used either on only one end of a call (to allow PCs to connect to regular phones), or on both ends of the call (to allow two regular phones to connect to each other — with the Internet as the long-distance carrier). We talk about PC-to-phone gateways in this chapter, and phone-to-phone gateways in Chapter 15.

Don't get the idea that gateways are only good for handling long distance phone calls, though — we talk about other uses of gateways in Chapters 16 and 17.

What's It All About?

It's about connecting the 'Net and the Public Switched Telephone Network (PSTN)! To make this connection, you need to hook up with a *gateway service provider.* These companies have decided that they can make money by offering users of Internet telephony programs the capability to call Grandma (or that friend in London), who doesn't have a PC, or maybe doesn't even know what one is.

When you use a gateway service provider to call a regular phone, the calls you make using Internet telephony software on your computer travel over the Internet to one of the service provider's gateways (preferably one near the person you are calling). The gateway then converts your Internet telephony packets to regular analog phone signals and sends them the rest of the way to your recipient via the PSTN.

The rates charged by these service providers usually vary widely, depending on the destination. If you call a location with a local gateway, your call is carried over the 'Net nearly the entire way, with correspondingly low rates. For locations that don't yet have a nearby gateway, a significant portion of the call is transferred over traditional long distance lines — with increased rates, as you would expect.

Many of the companies that are beginning to offer these services have a history as *callback* long-distance providers. Callback companies take advantage of the lower international rates that customers in the U.S. enjoy compared to the rest of the world. For example, a call from Amsterdam to New York may cost twice as much per minute as a call in the reverse direction. Callback systems allow overseas callers to make a brief (usually only a few seconds) call to a number in the U.S. This number then calls back to the overseas location (hence the name) and gives a second dialtone for the party to enter its destination phone number. Except for the first short, call, all of the calls originates in the U.S., so the rates are typically much lower than those for direct-dialed calls. In fact, the rates for callback calls between two countries — say Germany and England — are often cheaper when routed all the way through the U.S., so callback isn't just for calls into the U.S.

Some PC-to-phone gateway providers take advantage of this same concept in their calling plans by placing their gateways within the U.S. Calls from overseas are carried over the Internet to the U.S., and then transferred to the PSTN. In these schemes, the PSTN-billed portion of the call always originates in the U.S., so the call is almost always cheaper than a direct call.

What Do You Need to Use Gateways?

Using a PC-to-phone gateway is just like using any of the PC-to-PC programs we discuss in previous chapters. The hardware requirements are the same as for PC-to-PC 'Net telephony. In other words, you want to have:

- A relatively fast CPU (at least a 486 PC or a Power Mac)
- Full-duplex sound support
- A 28.8 Kbps or faster Internet connection

The software requirements vary depending upon which service you want to use. Some service providers — for example, Net2Phone — give you free, all-in-one client software use, while others, such as GXC, use an add-on program that works in conjunction with a commercial Internet telephony program like Internet Phone.

You also need an account number and a PIN (Personal Identification Number — like you use for your ATM card). Gateway services aren't free — to make calls like this, you've got to pay someone. Most gateway service providers (and not many exist yet) allow you to try out their service by making a few free calls to toll-free numbers. But after the free trials, you have to use your PIN to make calls, and you get billed for them like you do regular long-distance calls — although you'll probably get billed less than you would for regular long distance.

An interesting development in this area is the recent announcement of H.323-compatible gateways. Microsoft's NetMeeting already has the capability of dialing through an H.323 gateway to a regular phone — though we don't know of any service providers who currently offer a service that supports this capability. For now, this capability is probably most useful in intranet environments (which we discuss in Chapter 16).

Who Offers Gateway Services?

Only a few companies currently offer PC-to-phone gateway services. Although many companies have expressed an interest in this brand-new market, just a handful have actually started delivering services. We expect that more companies will soon offer these types of services, but many new entrants to the Internet telephony gateway market will likely skip the PC-to-phone business and head straight into the phone-to-phone over the Internet marketplace that we talk about in the next chapter.

IDT's Net2Phone

With its Net2Phone service, IDT — an Internet and long-distance telecommunications company based in New Jersey — is one of the first companies to offer PC-to-phone gateway services. With Net2Phone, you can call any phone you'd like from your Internet-connected PC. The rates for calls are based on the country of destination — in other words, your PC could be a town away from IDT's offices in Hackensack, or 10,000 miles away in Asia, and the cost of a call to the same number would be identical. This feature makes Net2Phone especially attractive for people overseas — folks whose nationalized telephone companies charge them an arm and a leg for international calls.

Net2Phone software is available for both PCs and Macs, and you can get a free download from IDT's Web site (`http://www.net2phone.com`). The first time you run the software (shown in Figure 14-1), you choose a PIN number. You can then try it out by calling toll-free (800 or 888) numbers, and if you want to use it for regular calls, purchase a prepaid calling card from IDT.

Figure 14-1: IDT's Net2Phone software running on Windows 95.

Global Exchange Carrier (GXC)

GXC is a Virginia-based start-up company dedicated to offering PC-to-phone 'Net telephony services. Using the Gx-Phone, a piece of software (shown in Figure 14-2) that works in conjunction with the PC versions of VocalTec Internet Phone, you can call regular telephones throughout the U.S. and in a growing number of international locations.

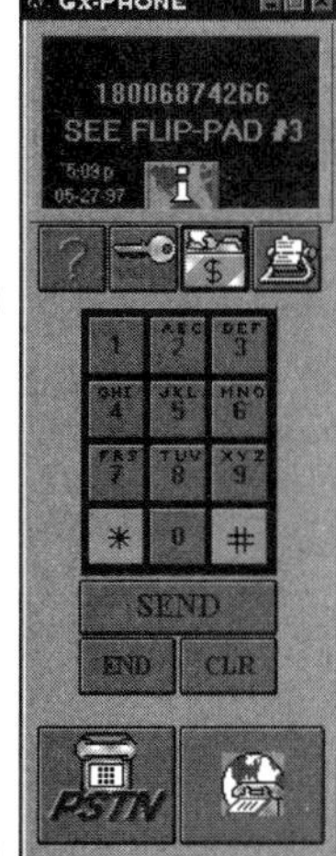

Figure 14-2: GXC's software works in conjunction with VocalTec's Internet Phone.

If you have Internet Phone, all you need to do to try out GXC is to download the Gx-Phone from their Web site at `http://www.gxc.com`. You can try out the phone for free by calling toll-free numbers. If you like it, the GXC Web site offers a secure form that enables you to establish an account. The Web site also lists rate information to various countries.

Killer Applications: Commerce and Call Centers

Although pay services that offer you the capability to call any phone from your PC will become more prominent and available in the near future, the really hot ticket for PC-to-phone Internet telephony is in the field of electronic commerce. In case you're not familiar with the term, *electronic commerce* basically means buying and selling things via electronic means — especially, in this case, over the 'Net.

Electronic commerce via the telephone is a huge market — just look at the vast amount of catalog shopping people do every year. Chances are good that you've spent some time leafing through a big fat catalog, marking down things you want and then calling an 800 number to place your order. We sure have. Well, imagine doing that same kind of shopping on the Web. You could point your browser over to the Web page of your favorite catalog, search through it until you find what you want to buy, and then click on a link to call a sales representative. This kind of technology is available now, and some big companies like Rockwell, Lucent Technologies, and MCI, as well as some others that you may not have heard of, like Dialogic and Netspeak, are working hard to bring it to you. The general idea is to take the *call centers,* or phone answering centers, that many companies use for sales or customer service, and hook them into the 'Net.

To do this, a company that wants to sell you something (or provide a service) places Internet telephony software links on its Web pages. Before the development of gateways, a company doing business on the Internet could only include a "call you back" link on its page — this type of link basically gives you the option of entering your name and phone number and then waiting for a representative to get that information and call you back. This isn't very satisfying if you want your question answered right now — and it's a real pain in the rear if you have only one phone line, because you have to disconnect your Internet connection, stop browsing the page, and wait for the phone to ring.

A call center gateway takes care of these problems. You can stay on the Web page, keep your only phone line tied up with your Internet connection, and talk to a company representative right away. The companies developing these systems are even designing features that will allow World Wide Web URLs (Uniform Resource Locators — Internet addresses, in other words) to be sent to your browser while you're on the phone with the company rep. So the sales rep, for instance, could automatically send you to a page with more information about the product in which you're interested. The gateway connects your Internet phone right into the existing call center phone lines, so a company can easily get this kind of system up and running.

The gateways also can handle calls that you make to a company via regular phone lines, sending the calls to representatives who are connected to the call center over the 'Net. This is basically the same concept turned on its head — instead of allowing you to call in over the 'Net, a company can put its phone operators just about anywhere in the world and distribute all incoming calls over the Internet. We read about one company in Germany that's already doing this — after-hours calls to their customer service line are actually forwarded to German-speaking operators in Florida over the Internet. In fact, as this technology is more widely distributed, you could end up making calls over the 'Net from your regular telephone without even knowing it!

Chapter 15

Telephone-to-Telephone Gateway Services

In This Chapter

- Is this stuff for real?
- How soon will it be available?
- What's holding it back?
- Will it be free?

Calling from a PC to a regular telephone is a pretty neat trick — and in the case of Web commerce and call centers, we think you'll see a lot of these types of calls in the future. But if the gateway can eliminate the need for a PC and an Internet connection on one end of the call, why not take things a step further and eliminate the PCs on both ends?

In other words, why not use the 'Net as the carrier for long-distance, phone-to-phone communications? As gateways improve, telecommunications companies — both established and start-ups — are working hard to figure out how they can offer this service (and compete with traditional telephone long-distance carriers).

As we write this, phone-to-phone Internet calling services are more of a coming attraction than the main feature. A few companies offer services — usually in very specialized markets (often quite geographically restricted — like calling to one particular country). No company has built a worldwide (or even close to worldwide) network of gateways that would allow you to call anywhere you want over the Internet. So you probably won't be trading in your AT&T or MCI long-distance account anytime soon in favor of Internet telephony.

Having given that warning, we can say that phone-to-phone Internet telephony (we sure wish we had a catchy nickname or acronym for that, because it's a mouthful) is definitely coming. A number of "Next-Generation Telcos," as they're often called, have announced plans to offer these kinds of services — a few have even announced worldwide networks. They won't find themselves alone in this market, however. According to gateway vendors we've talked to, a surprisingly large number of traditional telephone companies are testing out gateways, figuring, we guess, that if someone is going to sell Internet long distance as an alternative to traditional long distance, they may as well be the ones to do it.

What's the Deal?

We don't spend much time talking about "how to" in this chapter, because Internet telephony in this guise is really quite invisible to the user. But in the never-ending pursuit of completeness, we do give you some step-by-step instructions:

1. **Pick up the telephone and dial.**
2. **When the other party answers, talk normally.**

That's it. Really. At least that's how future services are being envisioned. You'll make long distance calls just as you do today, and you won't know or care that they go over the Internet — that is, until you get your lower phone bill at the end of the month.

In the meantime, as the first services begin to come to market they will probably be offered to customers as an adjunct, not a replacement, to their current long-distance service. You may have some additional access numbers to punch in on the phone, or perhaps a PIN number. However, phone-to-phone Internet telephony won't be any harder than other traditional long distance procedures such as calling cards or dialing in access codes to get your preferred carrier at a payphone. Easy as pie.

What Kinds of Services Will Be Available?

As we mention earlier in this chapter, the endgame (we love using catchy consulting terms like that in our writing) of all this gateway development and deployment is the capability for alternative long-distance providers to offer ubiquitous (or nearly ubiquitous — roughly half of the world's population has never even used a phone, so we're still a ways off from ubiquity), easy to use, direct-dial-style long-distance calls to and from PSTN telephones over the Internet. Sounds pretty neat, and it is.

In the interim, we think you'll find phone-to-phone Internet telephony available to you in some slightly less ambitious (but still useful) ways. Here are a few examples:

- **Calling cards:** At least one company already offers a calling card service for international calls. You simply dial into an access number and use your PIN to place the call.
- **Combined networks with least-cost routing:** Because no one yet offers a global network of gateways that enable you to call everywhere you can with traditional long distance, companies may begin to offer a service that combines both networks (the PSTN and the Internet), and automatically routes your call over one or the other, based on availability of network services and cost.
- **Geographically specific calling zones:** Until widespread deployment of gateways can be accomplished, service providers may specialize on a certain geographic area. For example, you may use your normal telephone service providers for 90 percent of your calls; but for the 10 percent of your calls that go to a specific area, you dial a special access number to place calls over your Next-Generation Telco.
- **Focus on international calling:** Domestic long-distance rates are relatively cheap, and may be coming down even more in the near future due to some recent Federal Communications Commission rulings. Phone bills pile up most on international calls — which can easily cost 10–15 times more per minute than domestic calls. No wonder then that most Next-Generation Telcos are focusing on this market and not worrying as much about domestic calling.

Who Offers (or Has Announced) These Services?

As you may expect with such a new technology, the equipment vendors are ahead of the service providers. The technology is established and proven, and the service providers are figuring out how to design and bill for the services. After they work out those issues, these kinds of services will be widely available. For now, however, we're looking at the birth of such offerings.

Like the PC-to-phone gateway service providers that we mention in Chapter 14, many of the initial entrants in this marketplace are currently selling callback services. Here are a few companies that will soon offer phone-to-phone calling over the 'Net:

- **USA Global Link:** A large, discount international telecommunications company specializing in callback services, USA Global Link recently announced that it is launching a worldwide phone-to-phone service called Global Internetwork. This system will allow you to make voice and fax calls anywhere in the world over a network that combines the Internet and private data lines, which the company plans to lease or own. You can find more information about this forthcoming service on the Web, at http://www.usagl.com/Internetwork/index.htm.

- **IDT Net2Phone Direct:** If you read the preceding chapter, you may remember Net2Phone, a PC-to-telephone service offered by IDT, an Internet and telecommunications company. IDT recently announced that it is launching a telephone-to-telephone service called Net2Phone Direct. Although the details are still sketchy, Net2Phone Direct will use local or toll-free access numbers to set up calls to more than 100 countries. Keep your eye on http://www.net2phone.com for the latest information.
- **Delta Three:** An Israeli company (a whole lot of companies in the Internet telephony field are from Israel, in fact), Delta Three specializes in selling gateway hardware for corporate use, but they also offer a calling card for calls from the U.S. to Israel. With this card, you call a toll-free access number, enter your PIN and the number you're calling, and your call is carried from the U.S. to Israel over the 'Net and then sent out to its destination via the PSTN. Delta Three is expanding its network, so you should soon be able to make Internet calling card calls to other locations throughout the world. You can find out the latest about Delta Three at its Web site, http://www.deltathree.com.

You'll soon see plenty of other companies offering these kinds of services.

What's Holding Back Development?

So telephone-to-telephone calls over the Internet are not a pipe dream — they really work, and people are really selling them. That doesn't mean we're ready to abandon our standard telephone service. Here are a few of the problems that must be solved before that day comes (if it ever comes):

- **Capacity of the 'Net:** The companies that provide the backbone of the Internet are spending lots of money every year increasing the size and the capacity of the fiber optic cables and associated hardware that make up the Internet. Users of Internet services, in turn, spend lots of time and effort finding ways to use that extra bandwidth. The bottom line: Capacity barely keeps up with demand as it is — dumping even a moderate chunk of the overall telephone network traffic onto the 'Net would cause things to grind to a screeching halt.
- **Compatibility:** Most gateways currently use proprietary techniques to send calls over the Internet. That's fine in a corporate intranet environment (see Chapter 16 to find out more about the use of gateways in corporate intranets), but could make it difficult to connect calls when

you venture out of this cozy environment. One way to make Internet telephony grow faster is by allowing Next-Generation Telcos to connect their networks together to expand the potential calling area. This will only work if their gateways are compatible — and many are not. Fortunately, many gateway vendors are beginning to offer or develop H.323-compatible gateways.

Interconnection: Another issue that remains only tenuously solved involves how to pay for use of someone else's Internet backbone network. Traditional telephone companies have elaborate "settlement" agreements in place with each other, to ensure interconnection and proper billing for carrying each other's traffic. In the Internet world, these kinds of agreements — often called *peering* — are often on shaky ground. Some backbone network providers end up carrying a lot more of someone else's traffic than they get carried in return. As you can imagine, this doesn't make them happy, and will continue to be a problem until someone devises a system of tracking and billing for traffic on the 'Net.

- **Regulatory issues:** Despite a decade or so of deregulation, telecommunications providers around the world are among the most heavily regulated and governed of all the world's industries. One of the reasons why Internet telephony is so much cheaper than traditional telephony is that in many cases, Internet providers are less regulated than traditional providers, and are not required to pay many of the access charges and fees that traditional providers pay. Our crystal ball isn't calibrated sharply enough to tell you what the future holds in this realm, but as more and more telephone traffic is carried over the Internet, you can bet that regulators (and those people who want to influence their decisions) will be watching closely.
- **Cost of deployment:** It's not too expensive to buy a couple of gateways and some reasonably high speed Internet access to connect telephones in two specific locales. That's why many companies are looking at Internet telephony as a good way to carry the phone calls between their own offices. The picture changes when you try to figure out the costs of deploying gateways all over the place — across an entire country or even globally — in an effort to compete head on with the traditional long-distance carriers. The cost and complexity of setting up a service of this kind is enormous — AT&T, Sprint, and MCI have good reason to employ tens of thousands of employees each.
- **Directory services:** PSTN telephone companies have a sophisticated and extremely reliable way of figuring out how to route and bill calls across their own networks, and more important, onto other people's networks. At this point in the development of phone-to-phone Internet telephony, no truly compatible directory service system exists — especially when you take into account the possibility of a call crossing from one Next-Generation Telco's network to another before being switched back to the PSTN.

The Real Goal: Convergence

Bear with us for a moment, while we talk like the visionary consultants that we are. (Really — we are! Trust us.) We don't think that Internet telephony is going to replace standard telephony any time soon. We do believe, though, that it will continue to grow and work better. The real goal here is convergence. Now *that's* a great consultant word for you.

Anyway, our point is that Internet telephony and traditional telephony can coexist. The real promise of gateway products is that they give service providers the option to route your calls over whichever network is more appropriate for the call at that time. The Web call centers we mention in the previous chapter offer a great example of that capability. You can initiate the call on the 'Net from a Web page. On the far end, the call gets routed to a sales agent or customer service rep over either the Internet or the PSTN, whichever makes most sense for that particular call. The point of convergence is this: You don't really care how your call is routed, as long as it goes to the right place at the right time and works well.

The price of quality

Ultimately, you will pay more for the higher quality service levels that will make the reliability and quality of Internet telephony competitive with existing phone services. Right now, no easy — much less elegant — method exists for prioritizing traffic in the free-for-all contention for bandwidth that characterizes the Internet, so your service provider can offer you only one service level. That will change in the future, when your service provider will be able to guarantee access to bandwidth — at least over your provider's network — for an additional fee. *Industrial-grade* services will be possible over at least some portions of the Internet, but expect to pay more for them.

The large service providers who provide Internet bandwidth are all rapidly expanding their networks — MCI, for example, is working on building its own worldwide Internet backbone. Service providers who can send most of their traffic over a single Internet provider like MCI can potentially bypass many of the chokepoints on the 'Net and give you a better quality of service. Adoption of new protocols — such as the RSVP bandwidth reservation protocol — will also help ensure quality. The downside of these types of technology, from a cost standpoint, is that they make the Internet more like the PSTN — where you end up "owning" a chunk of the network for the duration of your call. This approach can give you a guaranteed level of quality, but you have to pay for it.

Chapter 16

Using Internet Telephony in Your Corporate Intranet

In This Chapter

- What the heck is an intranet?
- Why should I make phone calls over my intranet?
- How can I use Internet telephony software on an intranet?
- Where do gateways fit into my intranet?

In the two previous chapters, we talk about some of the emerging ways that you can put Internet telephony to work with phones on the PSTN (Public Switched Telephone Network). Notice that we say "emerging." Although the capability for making calls to regular phones over the Internet is exciting, money-saving, and sometimes very useful, it's not really ready for prime time, yet. We believe that it will be, and soon, but at present the technology and the infrastructure are still being developed.

The place where this technology really works best — and makes a lot of sense — is within the confines of a corporate or enterprise intranet. As the Internet becomes almost ubiquitous, more and more companies are carving out their own little chunk of it in the form of an *intranet* — a private TCP/IP network running throughout and between offices, connected to the 'Net, but protected by firewalls that keep the bad guys out.

An increasing number of service providers offer services you can use to build and manage your intranet. Some ISPs now offer IP-based VPN (Virtual Private Network — basically a private data network carried over shared data lines) services using encrypted links over the Internet, or using separate connections to their backbones. IXCs (Interexchange Carriers, like AT&T or MCI) and local telephone companies are offering similar services over their networks.

The range of service providers announcing these services includes

- A growing list of ISPs, such as UUNET and PSI
- IXCs, including Sprint and MCI
- Local telephone companies, including Pacific Bell and US West

And many more companies appear poised to enter the fray.

Why Use Internet Telephony on Your Intranet?

If you were to take a sampling of Internet users and ask them about Internet telephony, you'd probably get one of two responses:

- Huh? What the heck is that?
- Ummm . . . Gimme a second . . . Oh yeah! Long distance for free!

We certainly hope that everyone who gives that first response picks up a copy of our book. As for the second response, we want to point out that even though the "free" part is a bit overplayed and oversimplified, Internet telephony can save you some money. And if your business is anything like ours, you appreciate that.

Low cost isn't the only reason to utilize Internet telephony on your intranet, but it sure helps when you try to convince the boss.

The bottom line is the bottom line

Say that your company has a main office in California, a branch in New York, another branch in Massachusetts, and an overseas office in London (everyone uses London for these types of examples — and we don't want to break with tradition). All of your computers are networked together via an intranet, which you lease from a big ISP. If you need to communicate with someone in the London branch, you send them an e-mail message. Fine, we love e-mail, too. But what if you need to speak with that person RIGHT NOW? Of course, you pick up the phone, dial the correct extension, and — bam — you're talking on the phone. At the end of the month, accounting pays the bill to the long-distance company. Ouch!

What about that intranet? Even though you probably pay a fixed amount each month for it, you may be using only 75 percent of its capacity. Why not take advantage of that extra 25 percent (which you pay for, regardless of whether you use it) and route that call over the intranet? We can't think of a good reason why you shouldn't do so.

Productivity ain't too shabby, either

Using Internet telephony on your company intranet offers more benefits than just saving money on long distance. Considering the increased productivity that some Internet telephony technologies offer, you may find it cost-effective to increase the capacity of your intranet to accommodate additional uses. Providers of intranet services certainly hope so!

How can you be more productive with 'Net telephony (and here we mean intranet telephony)? Here are some ideas:

- **Videoconference:** You can't always be in the same room as the people you're meeting with, but sometimes seeing them is just as good. Use of a product that allows multi-party videoconferencing, such as CU-SeeMe, gives you that capability.
- **Collaborate:** Consumer users of Internet telephony software probably scratch their heads and wonder why software developers bother to include all of those fancy whiteboards and such in their favorite programs. Add work to the equation, though, and the capabilities for sharing drawings or diagrams, transferring files, or even sharing applications suddenly make a lot of sense.
- **Plug in remote users or telecommuters:** Not only can you use a videocamera to see if your favorite telecommuters bothered to get out of their bathrobes today, you can also make it easier for them to communicate and work with those poor souls who still have to drag themselves into the office.

As you dig deeper into intranet telephony, we have no doubt that you'll find even more ways that it can make your life easier and allow you to work more efficiently.

Internet telephony just plain works better on an intranet

Maybe you've tried out some Internet telephony programs, talking from computer to computer over the 'Net, and thought to yourself, "Hey, this is pretty neat, but the delays are annoying . . . and the sound quality isn't so great." So you write it off as a neat toy for certified Net-geeks, or something you may recommend to your college-aged kids so they can call friends over summer break without running up your long distance bill. But you really don't think that it works well enough to be a serious business tool.

Well, we hate to disagree with you, our valued reader, but we do ask you to re-evaluate this technology in terms of your intranet. Internet telephony (maybe we should just call it *Voice over IP* in this context, because the 'Net is only peripherally involved) can work a lot better over an intranet, for a variety of reasons:

- **Bandwidth:** Chances are good that the computers attached to your intranet are plugged into a 10- or 100-Mbps Ethernet network, with your company's offices connected by some big, fat (or should we say fast?) multimegabit digital line. You probably use a lot of the bandwidth for other tasks — like database access, file transfers, and e-mail — and we don't recommend that you become profligate with it. However, you probably have enough left over to easily accommodate the requirements of IP audio, collaboration, or even video. You'll almost certainly be better off than someone dialing into an oversubscribed and underresourced ISP somewhere.
- **Latency:** When a 'Net call is routed over the public Internet, it can take a long and varied route between two points. Your voice or video packets may end up bouncing between a dozen or more routers before they get where you want them. Each time they hit a router, a certain amount of latency — or delay — is added into the equation, because the router has to figure out where to send the packet next. Add it all up, and you could end up with nearly a second of delay — which is certainly not amenable to a natural-sounding conversation. On your intranet, however, your packets probably hit only a few routers. And that decreased latency, combined with the increased bandwidth, makes for a good sounding — or looking — call.
- **A single, end-to-end provider:** You probably obtain your intranet as a service from a reputable, well-established service provider, maybe even one of the major Internet backbone providers like MCI or UUNet. In all likelihood, these companies can give you the increased bandwidth and lower latency that we mention — in fact, they may guarantee it. Additionally, you won't have to worry about having your packets transferred from provider to provider — a notorious source of delays and bottlenecks, if not outright failures to connect, on the public Internet.

Your mileage may vary, of course. How much better calls and conferences will work on your intranet depends on a variety of factors. In most cases, though, we'd be willing to bet that you will enjoy surprisingly good quality.

An Easy First Step: Using Client Software

Before you go whole hog and start buying gateways and an extra T3 line, you should probably get your feet wet by just testing out some client software programs (in other words, the kinds of programs we discuss in Part II of this book). Although some of the proprietary (non-H.323-compatible) programs certainly have their own merits, your best starting place may be with one of the telephony programs included with your browser software. We'd be quite surprised if you weren't already using either Netscape or Microsoft Internet Explorer, and those packages include full-featured, H.323-compatible telephony programs.

We encourage you to see what some of the other software programs have to offer as well. Perhaps you need videoconferencing that allows multiple parties to connect at once. If so, you need to go beyond the two programs we've just mentioned. Trial versions of almost every single program made are freely available over the 'Net, so you have no reason (except time) not to check out a whole bunch of them before you standardize on one or another.

While evaluating software for your intranet, consider the following questions:

- **How much does it cost?** This is really a no-brainer, but don't rule out a program just because it isn't free. It may well be worth the expense if it does exactly what you need.
- **How's the performance?** Another obvious one, but performance really does vary among different programs.
- **What kind of impact does it have on the intranet?** Even standards-based programs have various codecs available (in fact, the capability to plug in additional codecs is part of the standard), so if bandwidth is tight on your intranet, perhaps one particular program has a low bitrate codec that provides acceptable quality for you.
- **Is it easy to use?** Ease of use involves more than just the program's user interface. Think about how you place calls — do you to enter an IP address? Does the program have a good built-in phonebook?

Note: Some of the programs that use directory services to help resolve IP addresses for making calls offer a directory server program you can run on the intranet. (And other programs plan to offer this capability soon.) This feature can make the calling process a whole lot easier — especially as you add and remove people — and keeps your directory listings safe behind the firewall, instead of out on the 'Net.

If you find that a proprietary software program is best for your uses, try to find out whether an upgrade path exists that will eventually lead you to H.323 compatibility. Most programs are going this way, but it would be a shame to get a bunch of software installed and people trained, only to find that your needs include compatibility (perhaps if you add some gateways) and your existing base of software is now a dead end.

More Complicated, But More Sophisticated: Using Gateways

After you get your feet wet with some client software programs (as we describe in the preceding section), the next logical step involves moving your regular, interoffice telephone and fax (yes, fax!) traffic onto the intranet. This step — potentially your biggest money saver — is both exciting and a bit scary. It's exciting because of the following:

- The cost savings are tremendous, especially if you have international offices or a lot of interoffice fax traffic.
- The learning curve for end users is just about nonexistent — at most they probably have to learn a new dialing sequence.
- With the exception of the gateways themselves, you already have all of the equipment you need — no worrying about CPU speed or sound cards in all those phones on all those desks.

It's also a bit scary because of the following:

- You have to put some money down up front to buy the gateways.
- You may end up putting a lot more traffic on your intranet — this one always makes the network administrator types all sweaty.
- Your network folks have a learning curve to climb, if only while getting everything set up.

So taking this plunge is a bit scarier, because the diving board's a bit higher, but the water may end up feeling really nice. You have to spend some time seriously evaluating your needs, capacities, and costs before you make a move here (maybe you'll even end up hiring a telecommunications consulting company), but we think this step makes sense for a lot of companies.

We provide a listing of companies that offer gateway products in Appendix B. They (both the companies and the gateways themselves) come in all sizes and shapes — from simple four-line models based on Windows NT computers to huge carrier-grade (in other words, suitable for a phone company) models. We're willing to bet that your needs fall somewhere within that range.

Regardless of size or manufacturer, all gateways have a few elements in common:

- An interface with your intranet network
- Interfaces with your regular telephone system
- Hardware, specifically DSPs (remember them from Chapter 2?), to perform the digitization and packetization of your voice — and to undo it on the other end
- Software, to control the hardware, and to allow you to configure and set up the gateway

That's it! Well, the nuts and bolts are quite complicated, but the general concept is quite simple.

When you evaluate your choices of gateways, you need to consider some of the same points as you do for evaluating client software. How does it perform, in terms of sound quality? Does it use efficient codecs that won't overload your intranet? Does it recognize calls from H.323 software on PCs, or does PC-to-phone calling require proprietary software?

Here are some gateway-specific points to consider:

- How well does it handle fax traffic? (Faxing can be surprisingly difficult for some gateways.)
- Are faxes sent in real time, or are they stored and forwarded?
- What kind of phone interfaces are offered?
- What kind of special dialing is required on your phones to make it work?
- How expandable is the unit as your needs grow?

We're pretty sure that you'll have a friendly salesperson looking out for your best interests and helping you to answer these questions. Just don't forget to ask them.

Alternatives to buying your own gateway

Although buying a gateway and plugging it into your intranet is the main way a company can put intranet telephony to work, we think it's just a matter of time before service providers start to offer this as a value-added service to their current customers. This type of service has some potential benefits for users. For example, you could offload some of the set-up and configuration work (or all of it) to someone else.

Even more significantly, the service provider could show off the capabilities of its network and offer some sort of enhanced or guaranteed quality service, perhaps using RSVP. (We mention RSVP in Chapter 2 — it's a protocol for reserving bandwidth on a TCP/IP network.) You may end up paying more for this type of approach than if you did it yourself, but we'd bet that the service providers wouldn't do it if it wasn't going to save you money compared to your current calling method.

Chapter 17

Using the Internet for Faxing

In This Chapter

- Just the fax, ma'am
- Saving money
- Understanding how fax via the 'Net works
- Finding a fax via the 'Net service provider

In this age of electronic mail and file transfer, you'd think that fax might be heading toward obsolescence. Well, you could believe that, but all indications point to continued, rapid growth in the use of facsimile machines and other means of sending fax transmissions. A huge chunk of all long-distance traffic comes from fax machines.

In fact, a famous study (well, famous in the telecommunications industry anyway) conducted a few years ago concluded that fax machine use accounted for at least 40 percent of corporate long-distance bills. That's a pretty big chunk of money, by anyone's count.

In particular, the fax is often used for international communications, especially for cases in which time zone differences and language barriers make talking on the phone impractical. One study we read said that more than 70 percent of the phone calls between the U.S. and Japan are made between fax machines!

Why are we telling you all of this? To show you that companies (and individuals) spend a lot of money each year sending faxes. So, we believe you have a compelling reason to look into sending faxes via the Internet as a cost-saving measure.

Why Does It Make Sense to Fax on the 'Net?

We can think of several reasons why it makes sense to fax over the Internet (or an intranet) instead of using the PSTN — for example:

- **Fax really is data:** When you get right down to the bottom line, a fax transmission is a data transmission. So why send it on an expensive voice network (the PSTN) when you can use a cheaper data network (the Internet)?
- **Fax isn't as delay-sensitive as voice or video:** When you have a phone conversation, or a videoconference, you expect only minimal delays in the transmission of your voice (or video) signals. When packets of data get lost or misrouted in a phone call, you end up with chunks of conversation lost, or an unacceptable delay while they are reassembled. Annoying, to say the least, and possibly bad enough to make the conversation untenable. With a fax, on the other hand, do you really care if it takes a few extra seconds to get transmitted because of delay? We don't, and you probably don't, either.

  Note: Although users typically don't care if a bit of a delay occurs during a fax transmission, fax machines sometimes do — they may see the delay as a lost connection and stop the fax transmission. Gateway manufacturers who want to send faxes in real time between fax machines have to design a way of "spoofing" the fax machine to ignore these delays.
- **Faxes often aren't time-sensitive:** Sure, sometimes you desperately need to receive a fax, and you camp out in front of the machine, ready to tear off the fax as soon as it's printed. But you probably also send (and receive) plenty of other faxes that don't need to be read right away. So why spend money on sending them in real time over the PSTN when you can send them in a non-real-time way (*store and forward,* as it's often called) over the 'Net?

Of course, the number-one, most compelling reason to send faxes over the Internet is because it's cheaper than sending them via the PSTN. And in a lot of ways, faxing over the Internet is even easier than sending real-time voice and video data over the 'Net. It's somewhat surprising, then, that voice and video really took off first, but we think fax over the 'Net — and the service providers who offer it — will grow by leaps and bounds in the next few years, perhaps even eclipsing voice telephony.

How Does Internet Faxing Work?

The key to Internet faxing — at least, what we mean when we talk about Internet faxing — is the gateway. In the past, some schemes sent faxes as e-mail and then printed them out on fax machines, but that's not what we're discussing here. The real action happens when a fax machine is involved on one or both ends of the call, and the fax is sent as a fax, not as plain text within an e-mail message.

Like voice Internet telephony through a gateway, Internet faxing can come in one of two flavors:

- **PC to fax machine (desktop faxing):** In this case, a PC connected to the 'Net uses special software to send a document (any document, from any program, for the most part) over the Internet and through a gateway to a fax machine that prints out the document. (This method works a lot like using the software that comes with many fax modems, but instead of dialing the fax machine on the other end directly over the PSTN, the Internet carries the fax.)
- **Fax machine to fax machine:** Like phone-to-phone calling over the 'Net, a fax-to-fax call goes through gateways at both ends of the call, with each fax machine connected to a regular phone line.

Internet faxes are transmitted over the network in one of two ways:

- **Real-time faxes:** Just like faxes between two machines (standalone fax machines or PCs with special fax software) on the PSTN, a "connection" is made, and the fax is printed out as the receiving machine gets it.
- **Store-and-forward faxes:** In this case, the sender's PC or fax machine transmits faxes to the gateway. The gateway receives the entire fax, saves it as a data file, and then sends the file (sort of like an e-mail message) to the receiving gateway. The receiving gateway converts the data file back to a fax transmission, and sends the complete fax to the receiving machine, which prints it.

How to Get Started with Internet Fax

The Internet faxing marketplace focuses primarily on business customers. This should come as no surprise — consumers are far less likely to use fax than businesses. Business users have two ways to get into Internet faxing:

- **Do it yourself:** For intracompany faxes, you can use many of the gateway products that you already may be considering for your intranet to send and receive faxes. (See Chapter 16 for more on this subject.) Typically, these gateways allow you to plug in the fax machines like you would telephones and go to it.
- **Buy an Internet-based fax service from a service provider:** This option saves you from having to buy and set up your own gateways, and typically affords you a much larger coverage area.

Note: Like the Next-Generation Telco service providers we discuss in Chapters 14 and 15, many Internet fax service providers specialize in geographically limited service areas. For example, some specialize in overseas faxes only — which is where you can save the most money compared to the PSTN.

As you can imagine, consumers don't really have the option of setting up their own gateways. We suppose you could do it, if you really wanted to, but we're not quite sure why you'd want to spend thousands of dollars for a gateway in the spare bedroom.

So consumers must purchase Internet fax service from a service provider. Note that we said "purchase" — although it's cheaper than PSTN faxing, no one gives away Internet faxing for free. We wouldn't be surprised, however, to see Internet service providers begin to offer 'Net faxing as a value-added service, as a means for differentiating themselves in their extremely competitive marketplace.

Some Service Providers

Only a handful of 'Net fax providers exist right now. Internet fax is all brand-new, state-of-the-art stuff. We believe you'll soon see many more companies move into this market, as they begin to realize its vast potential.

FaxSav

FaxSav is a global fax networking company that uses a combination of the Internet and its own private faxing network to deliver faxes from either fax machines or PCs to fax machines worldwide. Pricing depends, of course, upon destination, and faxing to places where FaxSav has an Internet gateway (they have them in 14 countries as we write this) is cheaper than to places where the fax ends up being carried on a traditional telephone network.

FaxSav uses an interesting piece of equipment for fax machine-originated faxes — a special dialer that automatically routes your call to either FaxSav's network or to your normal long distance provider, depending on the destination of the fax and what service options you have purchased from FaxSav.

FaxSav's desktop faxing system includes free software for Windows or Macintosh, and a free 10 page fax trial. You can download the software and find more information about the trial at `http://www.faxsav.com`.

NetCentric FaxStorm

NetCentric's FaxStorm service provides desktop faxing over the 'Net to fax machines throughout North America and indeed, throughout the world. Rates to international destinations vary depending on whether FaxStorm has a gateway in that particular country (you can check out the rates to various countries on its Web site, at `http://www.faxstorm.com`).

FaxSav has free Windows software and trial accounts available from the Web site. If you like the service, you can sign up for either a personal or a corporate account, depending on your situation and needs. You can also use the popular Delrina WinFax PRO software with a FaxStorm account.

Future providers

As we mention earlier in this chapter, we think fax over the Internet is going to be a fast-growing market — perhaps even eclipsing standard faxing in the next few years. We expect an explosion in the number of companies offering 'Net faxing services in the very near future.

We also expect that many ISPs will offer desktop Internet faxing as a low-cost option or as part of an enhanced service package — in fact a few, such as Netcom, have already announced their intention to do so.

ISPs who offer intranet services to companies will also begin to offer fax as a part of their service packages, just as they may begin to offer voice services. And just as many Next-Generation Telcos are popping up and beginning to offer voice over the Internet service, a host of companies will soon specialize in fax over the 'Net — along with Next-Generation Telcos who offer fax machine to fax machine service to go along with their telephone to telephone service.

Part IV

The Part of Tens: Hip Stuff from the Internet

"You know, I liked you a whole lot more on the Internet."

In this part . . .

Top ten lists galore! Look in this part of the book for top ten lists on all sorts of things about Internet telephony and video.

Find out where to go for more information and how to keep up to date on this fast-changing field. Read ways to improve how you look and sound on the 'Net and how to use Internet telephony and video to make your home and business more productive (and more fun!).

We also put on our crystal ball-reading glasses and give you a glimpse into the future of Internet telephony and video. It's not some pie-in-the-sky long term look either, but rather a look at some things that are happening in the here and now that will have huge effects on the 'Net in the next few years.

Chapter 18

Ten Ways You Can Use Internet Telephony and Video Now

In This Chapter

- Keep in touch
- Save money!
- See your kids/folks/grandkids/aunts/uncles (and even the beagle)
- Do the virtual office thing

We spend a great deal of time online and a great deal of time using 'Net telephony and videoconferencing programs. We hope that fact is obvious to you by now. We also spend a great deal of time talking to people we meet online, asking them questions and picking their brains. All for you.

So when we had to decide on ten ways that you can use these products now, the problem was more of picking only ten ways to do so!

We hope we included your favorite use in this list (unless your favorite use was something that you wouldn't want to tell your mother you were doing). You can do plenty more with this cool new technology, so consider this just our Top Ten list.

Talk to Your Friends

A large part of using any telephony product is using it to keep in touch. Unless you are a great deal younger than we are (and we're not that old), you remember the "Reach out and touch someone" ads that AT&T used to air. If you were talking to us on the Internet right now, Pat would probably sing it for you. (Thank your lucky stars you're not online right now.)

It may be corny, but it's true! An unbelievable number of people we bumped into on servers and reflectors and elsewhere told us stories about having a computer, being on the 'Net, and happening upon some telephony and video software. Next thing you know, they've got the whole family online, and everyone is talking to everyone else.

One group that seems to be taking advantage of this new capability is college kids. Many, if not most, college students have a multimedia PC sitting on their desks already and unlimited 'Net access through their school. Before you can say "The Duke Blue Devils are the best basketball team around," Mom and Dad and the best-friend-at-a-different-school are online, chatting away.

It's not just the young that are online, either. The elder generation is one of the fastest-growing, computer-owning, 'Netizen segments of all — the big gap in computer usage is actually being found in the middle. So we've run into dozens of grandparents who use the 'Net to talk with (and see) the grandkids. We even talked to one fellow who went out and plunked down a big wad of cash for a color camcorder to replace his QuickCam. He wanted only the best quality when his wife was online with her grandchildren — black and white just wouldn't do!

Save Some Big Bucks

Long-distance rates may be a great deal lower than they used to be, but they can't approach the near-zero cost level of Internet telephony — especially for international calls. We've talked to people all over the world — in places such as Western Europe, Scandinavia, and all over Asia. Imagine the bills you could ring up (no pun intended) on those calls.

Not only is calling these places by using the Internet much cheaper, sometimes it's even easier. We've talked to people who were in countries where it's actually easier to get Internet access locally than to try to get connected to an overseas long-distance carrier. The Internet is growing faster in many places than the PSTN (Public Switched Telephone Network).

Videoconference on the Cheap

Traditional videoconferencing equipment (the fancy dedicated stuff used by big corporations) is expensive to buy (maybe $1,000 or more per participant) and use (dedicated lines are required). Even ISDN-based desktop videoconferencing systems (such as Intel's ProShare) have a high entry cost plus expensive ISDN line charges for point-to-point transmission (local fees plus long distance).

Internet videoconferencing, on the other hand, makes use of what you've already got — a 'Net connection and a PC. All you need to do is spend a hundred bucks on a QuickCam and download some videoconferencing software. We're not going to tell you that the quality is as good as some $1,500-per-seat dedicated videoconferencing system, but it gets the job done; and the quality can only get better as time goes by and things such as access speed and compression improve.

Call Home from the Road

Most new laptops have multimedia capabilities (sound support, microphone, and speakers) built right in. When you're on the road, just dial into the local number, get online, and call home without paying those outrageous hotel long-distance rates.

So if you are a road warrior, choose an Internet Service Provider with POPs nationwide (most of the big ones, such as AT&T, Netcom, and PSINet, have them almost everywhere) so that you can log on at will. In fact, some of the major ISPs are starting to make deals with carriers and ISPs in other countries, so you can log onto your account from just about anywhere in the world. Or pick a carrier with toll-free access (but watch for potential extra charges).

Meet New People

If you are familiar with the 'Net or with online services such as AOL, you have no doubt seen the meeting places, such as chat rooms or Internet Relay Chat (IRC). Internet telephony is really the next step in this concept. Log onto an Internet Phone server or onto a public CU-SeeMe reflector, and you can bump into literally dozens of interesting people — talking about all sorts of subjects.

The server-based products are best for this sort of thing — Internet Phone, in particular, always has a big crowd of people looking to chat on its server. CU-SeeMe users are another friendly bunch, with Web pages devoted to screen shots of each other and many people holding real-life get-togethers to see each other without the camera.

Talk to Your Co-workers and Customers

'Net telephony and video products aren't just for having fun and goofing off — they can be serious business tools as well. At the recent Interop trade show in Las Vegas — the major get-together for us telecom geeks — the keynote speeches were transmitted around the show by using Enhanced CU-SeeMe!

As the virtual office concept turns from theory to reality, more and more businesses find themselves spread out across states (and even countries). The Internet and Internet telephony can help narrow these geographical distances.

Companies can use Internet voice products in one of two ways:

- To talk with people outside the company over the Internet. For example, a company may want to put a link to a 'Net telephone "number" on its Web page, so customers cruising the Web page can call into a live person for more information.
- For internal company communications on an Intranet (see the following sidebar). Why spend extra money on phone calls between offices when the offices are already connected on an Intranet? The employees already have computers sitting in front of them — use those computers to talk. Or even better, set up a gateway system, and use your regular phones and fax machines to communicate between offices.

What the heck is an Intranet?

Intranet is the buzzword of the year, for starters. An *Intranet* is basically a private chunk of the 'Net, cordoned off by passwords and firewalls, that is used just for an organization's internal use. The Intranet's look and feel is identical to that on the rest of the 'Net, with Web browsers ("Universal Client" is the term that Netscape employees are now required by corporate policy to use when talking about Netscape Navigator), TCP/IP, and all of that familiar Internet stuff.

Why are Intranets so hot? Because corporate information systems managers have been trying for years to move away from the old rigid and hierarchical *mainframe* way of looking at their corporate networks — sometimes with mixed results. The boom in the World Wide Web caught the IS managers' attention — the 'Net is the ultimate in decentralized networks — and they have begun (in a big way) to move toward applying the principles of an open, flexible Internet to their own internal needs.

Share a Document

Sharing of documents has been around for a long time — ever been to a meeting that didn't include some kind of photocopied agenda? The latest generation of Internet telephony and video programs puts this functionality on your desktop, no matter where you are in the world.

Forget faxing an agenda or a spreadsheet that needs to be reviewed. Just use the whiteboard function of Internet Phone or CU-SeeMe (among others) to share the information with your co-workers. You can all view the same document, make notes and comments on it, and after you are finished, save the marked-up document for further study. With NetMeeting you can even share applications — edit and make changes to a document on someone else's computer, over the 'Net!

At-home uses for this technology exist, too. Got a great picture of the dog wearing a goofy little hat? (We don't condone putting hats on dogs, by the way; but live and let live.) Scan that puppy into your hard drive and post it on the whiteboard while you are talking to your friends. If you are using CU-SeeMe, you can even see the look of shocked disbelief on their faces.

Link Up Your Web Site

Got your own home page? Make it interactive by putting a link to your Internet Phone address on it. Your friends don't need to search you out on the server or even put you in their speed-dial buttons. All that your friends need to do to call you is to go to your Web page and click the link to connect.

See and Hear Before You Buy

Streaming audio and video products can bring electronic commerce to life. Buying and selling over the Web is an increasingly important way of doing business — you can't watch more than ten minutes of television (or read more than ten pages of most magazines) without running into an advertisement that includes a Web page address.

That's great, but the still graphics and text of most Web pages don't really offer much more than a paper catalog does.

That's where streaming audio and video come in. Many companies are integrating audio or video descriptions of their products with secure (at least they claim that they are secure) credit-card ordering systems. So you can go to the Web page, read about the product, watch a video of it in action or listen to a description of it, and then place your order.

A good example of this is the Music Boulevard Web page (at `http://www.musicblvd.com/`), where you can browse through tons of available CDs and tapes, listen to clips of the ones that you're interested in, and then purchase them online. Fun and easy.

Listen to Your Hometown Radio Station

A growing number of radio stations are beginning to offer a live RealAudio simulcast of their programming over the 'Net. University of North Carolina at Chapel Hill fans, for example, can listen to their 'Heels getting creamed by the Duke Blue Devils on RealAudio at `http://www.goheels.com`. (Figured out that two of us are from Duke University yet?)

In fact, geographically dispersed niche markets, such as college alumni, are the perfect audience for such products — products that offer the kind of specialized focus that the once-much-acclaimed 500-channel cable TV of the future was supposed to offer.

One of the authors found a personal use for this when he was able to listen to his favorite station from home (`http://www.91x.com`) while working on this book 2,800 miles away. The 'Net really does make the world smaller.

Chapter 19

Ten Hot 'Net Sites Where Video Is Used Today

In This Chapter

- The news
- More news
- Even more news (sensing a pattern?)
- Movie clips from Finland
- Space shuttle flights
- Conferences and stuff
- Golf
- PBS — but no Bert and Ernie yet

Online video isn't just "coming soon," it's already here. Despite the fact that today's systems may not yet be ready for prime time, you can find some cool and fun (and even productive — if your desk is near the boss's office) examples of video on the 'Net today.

One of the most common uses of Internet video is news delivery. All three of the major networks have some sort of 'Net video presence — nascent though the technology is. Why is news a video killer application? We can think of four reasons (feel free to add your own in the margins):

- The networks all see a need to create some sort of Internet presence (they're not dumb, you know), and news is a quick and effective way to get people's attention.
- We hate to say it again, but the quality isn't that great yet. Ninety percent of TV news is watching a head talk, so the quality demands aren't as great as they would be for watching the blood-and-guts scenes on *ER*.

- A great many people work, and many of them don't have the luxury of flipping on the tube in the office when they want to catch the headlines. But they do have the Internet sitting right in front of them — so Internet video news providers can catch them right at their desks, in the middle of the day.
- The Internet is becoming an increasingly important place for people to turn to for news coverage. Chances are that your local newspaper has (or is developing) some sort of Web presence. And chances are also good that, if you're a 'Net hound like us, you probably breeze through c|net's technology news section or stop by The Weather Channel's Web page every time you're online. So adding video news to a Web page is a good way of getting your attention.

More than just news is out there, though; so if you've sworn off watching the news to protect your blood pressure, don't worry.

MSNBC Desktop Video

NBC used to offer its news on the 'Net by broadcasting a product known as NBC Pro desktop video. That was last year, however. Since then, NBC has teamed up with a small Seattle company known as Microsoft and formed a new cable and Web network called MSNBC. Not surprisingly, this is a forum for broadcasting a business-oriented news service. MSNBC uses Microsoft NetShow and Progressive Networks's RealAudio to give you live business news all day and night. To find the broadcast, go to `http://www.businessvideo.msnbc.com/`.

CBS Up to the Minute

CBS's *Up to the Minute* (UTTM) overnight news program airs from 2 to 5 a.m. every morning. We're not sure about you, but we prefer to be sound asleep during those hours. (Even deadlines can't keep us up then.) Luckily, you can go to the *Up to the Minute* Web Site at `http://uttm.com/` and watch a selection of UTTM segments during more reasonable hours. CBS uses the VDOLive system to deliver these segments on demand, so you day people can keep up and not sleep through the alarm clock.

Fox News

The newest of the major networks, Fox, is also new to the news business. Fox has certainly hit the ground running, though, and that includes a strong Internet presence. As a part of its launch, Fox is offering a 24 hour a day, live 'Net broadcast of all its news programming. Fox has decided to use Progressive Networks's new RealVideo system to provide this round-the-clock service. To catch the broadcast, all you need to do is fire up your browser and head to `http://www.foxnews.com/`. Once there, you'll find a link on the main page to view the live video (like the example in Figure 19-1) — so get watching!

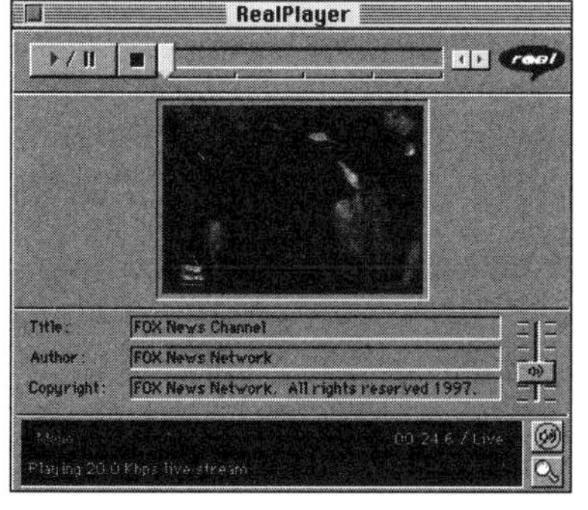

Figure 19-1: Catch the latest from Fox News on RealVideo.

KPIX TV

KPIX, a TV station in San Francisco, makes many of its news segments available on its Web site (at `http://www.kpix.com/video`) for playback on demand, using the VDOLive system. Go to the KPIX Web page to see the latest list of available videos. We watched Dan Fouts (the Hall of Fame quarterback) take a hike. (No pun intended; he really was hiking on a trail.)

Rocket Science

NASA broadcasts live pictures from Mission Control and the space shuttle during operations. You can log onto one of many reflectors (available in a multitude of places) by using CU-SeeMe, and catch all of the action live.

The pictures are not always great; but if you love space stuff, you don't want to miss NASA. One word of warning, though — make sure that you set up CU-SeeMe not to send video before you begin. Most of the reflectors are set up not to receive; but if you somehow slip through, you're going to feel pretty darn stupid (and maybe have a few dozen peeved people watching you stare at the picture). Enough said on that.

The reflectors you can use include:

- NASA's reflectors at 139.169.165.25, 139.88.27.43, or 128.158.1.154
- Kent State University's reflector at 131.123.5.1
- IITAPs reflector at 129.186.112.242

Finland Online!

For some reason, we are fascinated with reading (looking at, actually) the Finnish language — don't ask us why. We're not sure, either. Luckily for us then, Finland's telephone company, Telecom Finland, has launched an experiment in Internet broadcasting called MediaPilot. You can access MediaPilot through its Web site at `http://www.ml.tele.fi/`.

StreamWorks audio, video, and live (both audio and video) broadcasts of all sorts of things are available, ranging from children's shows in Finnish (we think that's what they were — we saw a bunch of kids talking to each other, and we couldn't tell what they were saying) to trailers from the latest Hollywood movies. We saw clips from movies that haven't even been released stateside yet, so MediaPilot is keeping on top of updating content. (*Pet Peeve:* Web sites that *never* get updated — boring!)

Random Tube

The Digitcom Multimedia Corporation, at `http://www.div-n.com/`, broadcasts various on-demand and live video streams, using a whole bunch of systems: StreamWorks, VDOLive, RealAudio, and VivoActive.

Among other things, you can use VDOLive to access the Video Jukebox, a collection of video demonstrations (and probably a whole bunch more, by the time you read this), or you can use StreamWorks to view live TV news and other programming. (Last time we were there, VH-1 was playing.)

Fore!

We aren't big golfers, but we know plenty of people who live and breathe the sport. We're including this one just for them: The Golf Channel.

Actually, we can appreciate that the Golf Channel is online (at `http://cbs.sportsline.com/u/golfchannel/`), because it gives those inveterate golfers an alternative place to be instead of tying up our cable-TV channel with the Masters from 17 years ago. (*Authors' Note:* Pat wants to assure his family that he is not referring to any of them in that previous statement; Danny says Pat is lying.)

You can log onto this page and, by using StreamWorks, watch on-demand feature clips and some special live events. Last time we were there, the Golf Channel had video clips (of Johnny Miller) and profiles (of Arnold Palmer) for your viewing pleasure. If either of those two names gets your blood rushing, point your browser, fire up your viewer, and enjoy!

Music Videos

These are not quite the type you may be used to. Boston's Channel 1 Internet Service Provider (at `http://www.channel1.com`) has launched a site called Graphic Audio (at `http://www.graphicaudio.com`) that offers clips of rock concerts, independent films, and other neat stuff in VDOLive format.

The content is presented in a plug-in format; so if you have Netscape 2.0 and the VDOLive plug-in installed on your machine, you can just go to the page and tune in some cool stuff.

PBS

It's not just Oscar the Grouch anymore. PBS (at `http://www.pbs.org`) has a streaming video service using the VDOLive player. (Streaming audio is also available using the RealAudio player.) There's not yet video for every one of the many PBS shows, but you'll probably find some for at least one of your favorites. Pretty neat stuff, and it sounds as if they are going to be putting a great deal more of their stuff online in the future.

Chapter 20

Ten (Plus One) Tricks to Sounding and Looking Good on the 'Net

In This Chapter

- Stupid bandwidth tricks
- Four ways to make friends with your microphone
- Big smile for the camera and four other video tips

You can't control every aspect of how you look and sound on the 'Net — your data flows across a whole bunch of wires, and cables, and switches, and routers, all of which make up your local Internet Service Provider (ISP) connection and the Internet backbone (and what happens there is anybody's guess).

You can, however, control your inputs to this (effective) mess. Remember GIGO — garbage in, garbage out. The best transmission method doesn't do you a bit of good if the audio and video signals you are sending into the computer are poor.

Following are some tips for optimizing what you send out. One costs you a few bucks, but the others are free — all can help.

Get a Bigger Pipe

This one is the tip that costs you, but it can probably bring the greatest improvement: Look into getting a faster connection into your ISP. If you are using a 14.4 Kbps modem, consider upgrading to a 33.6 Kbps model — the investment is usually pretty small (200 bucks tops), and we'd be surprised if your ISP didn't support at least this speed.

ISDN (Integrated Services Digital Network) can offer an even greater improvement (with a correspondingly larger bite out of your bank account every month). We are slightly hesitant about giving a blanket recommendation to ISDN, for two reasons:

- Prices can vary wildly, even within the same telco. Some carriers charge three or four times as much as others, and some have that much of a difference in price in different service areas. (One state in a telco's service area can be extremely inexpensive, while an adjacent state is outrageously priced.)
- Some new technologies are brewing on the horizon that blow ISDN away in terms of bandwidth and cost effectiveness (see Chapter 25).

If you are lucky enough to live in a low-priced ISDN service area, by all means consider it. Video, especially, benefits from the increased bandwidth; and everything you do on the 'Net, from FTPing files to opening graphically rich Web pages, goes faster.

While you're waiting for that ISDN line or new modem to show up, be smart about the bandwidth that you do have. Figure 20-1 shows the Network Statistics dialog box for a call we made with a friend who was performing an FTP download of a large file at the same time that we were talking. Not only were the stats bad, but the quality of the call was lousy, too!

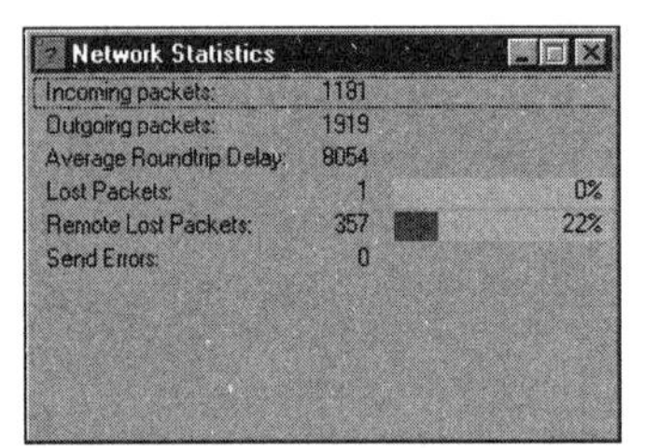

Figure 20-1: Don't try this at home — trying to talk on the phone while downloading a file.

Use Your Microphone Correctly

This tip is a multipart item, and most of this information is just good common sense. If you take our advice and do these things, you can probably see a substantial improvement in the quality of your voice transmissions for little or no money. Really!

Don't overdrive your mike

Probably the most common misconception in new users' heads is that they need to be right up on their microphone to talk into it. After all, music stars' lips brush right up against their microphones in concert, right? So if 'Net phoners want the same fidelity, they need to do the same with their computer microphones, right? Wrong!

We've talked to many first-timers, and the impression that we always get is of someone hunched up over their desk, with their mouth inches from the microphone saying, "Can you hear me?" (With the video programs, we actually see this — and it is not a pretty sight.)

You don't need to be that close. When you're that close to the microphone, your voice level is just way too high; and all you really hear on the other end is a fuzzed-out, distorted "Canttpptthpttyouhesshhhhhhhhhzzzttttt-mmeeettht."

What you need to do is put a little distance between you and the microphone. We don't suggest sitting across the room from it, but 12 to 18 inches is probably a good place to start. Maybe you have a microphone that can sit on top of your monitor — if so, put it there. If you have a freestanding desk microphone, try putting it behind your keyboard. Just don't put it where you find yourself leaning into it.

Spend a few minutes talking to yourself to get used to the microphone — use your computer's sound recorder program to record yourself speaking, and experiment with different microphone placements. If you are using a boom-type microphone or an in-the-ear model (such as a Jabra Earphone), follow the advice in those products' documentation. These kinds of microphones were designed to be closer to your head, and they usually give very good results.

Set the voice activation level correctly

Almost all the microphone products use some kind of automatic voice activation system, which detects when you are speaking and begins to transmit audio. As a general rule, you should set this control so that it is as low as possible without being activated by the background noise in the room. If you set this control too low, however — so that it does pick up this background noise — you send out packets all the time, which can either be annoying (in full duplex) or catastrophic (in half duplex — the other party can never say a word). On the other hand, if you set this control too high, you practically have to yell to activate your outgoing audio. In this case, you're likely to find that the first few words you speak never get transmitted. Needless to say, voice activation may take some tinkering to get it set just right.

Speak clearly and without long pauses

We don't need to tell you why you should speak clearly, but we must remind you to do so. Mumbling can be annoying in person, but believe us, it's ten times worse over the 'Net.

You should also avoid long pauses when you speak. If you wait too long between words, the chances are good that your program will stop transmitting audio. The party on the other end will notice this, think that you've finished speaking, and start to reply. Of course, by this time, you've begun to finish your thoughts, and you'll be talking over each other. This makes for a lot of "Oh, I'm sorry, go ahead" phrases shooting back and forth. Busy times on the 'Net (which may increase the lag time between speaking and being heard) only add to this potential problem.

Use a headphone

Full duplex is great — it can make a great difference in your perception of call quality. But if you have a set of speakers and a microphone sitting next to each other on your desk, you're probably going to get feedback. That is, the other person's voice goes out of your speakers and into your microphone and is sent back to that person. This situation can get confusing. It's like the folks who call into talk-radio shows and then don't turn down their radios — you end up listening to yourself.

To alleviate this, try using a headphone setup. We had great luck with the Jabra Earphone, but plenty of others are out there as well. Whatever headphone model you choose, it can make a big difference in the quality of your call.

Smile and be nice

This is what they tell telemarketers and people doing job interviews over the phone, and it's good advice in general. If you are sitting in front of your computer with a big frown on your face and a little black cloud floating over your head, you sound like that on the other end. Talking on the Internet is supposed to be fun (and it is!), so enjoy.

Also don't do any of the following things while talking on the 'Net:

- Eat a cheeseburger.
- Drink orange juice after brushing your teeth.
- Take bad-tasting medicine.
- Chew an entire pack of bubble gum.

Say Cheese!

Unless you are a TV talk-show host when you're not online, you probably want to spend some time perfecting your on-camera presence. (Pat's still working on his, though Danny is a natural ham — the jury's still out on Rebecca.) You need to remember or look out for several things.

Lighten up

TV news crews carry a bunch of klieg lights with them on location for a reason — you look much better on camera with some decent lighting. Here are some do's and don'ts:

- Do use some additional light source to light up your face.
- Don't think that the overhead lights in your office are good enough, especially if they are fluorescent. (For some reason, they really stink on camera.)
- Do put your light source behind the camera, where it reflects off your face, not off the camera's lens.
- Don't sit in front of the window — the sunlight can make you look like a silhouette, and your facial features can disappear from view.
- Above all, watch for shadows, because they make you look haggard and old.

Look at the camera

Camera placement and your behavior on camera can greatly influence the success of your video presentation. You'd be surprised at the number of foreheads you can see at any given time during a videoconference — it seems as if everyone is looking at their shoes.

Unless you are planning on showing off your new hair style, your facial expressions are the key elements that video can add to a conference. If you just want to see what everyone looks like, you can look at a still picture. But if you want to get a feel for how your new proposal is received, seeing the smile or grimace on your boss's face is important.

Position the camera intelligently

The QuickCam fits nicely onto the top of your monitor, but conference participants spend all their time watching everyone else on the screen — so all you see is the top of their heads. You may want to consider purchasing an inexpensive tripod (a receptor for a tripod is built into the bottom of the QuickCam) and placing the camera at a level even with the middle of your screen. In fact, Nokia, the Finnish electronics maker, recently released a multimedia monitor with a camera built into the ***bottom*** of the monitor casing, because most people look down a bit to see their monitor.

Take notes judiciously

If you are reading or taking notes, try to put them between yourself and the camera and not off to the side. Don't forget to look up once in a while. If you are giving a presentation, you may want to have someone hold up cue cards (like TV newscasters use) behind the camera so that you're not looking down all of the time.

Try to hold still

We're not suggesting that you hold your breath all the time; but if you drink 12 cups of coffee before you start the conference (and spend the whole time sliding your chair from one side of your desk to the other), your picture is going to be lousy. Remember that video compression algorithms save bandwidth by sending only the part of the picture that has changed since the last frame — so the less you move around, the better your picture is.

Chapter 21

Ten (or So) Good FAQs and Mailing Lists to Check Out

In This Chapter

- VON (Voice on the 'Net) mailing list
- Internet Phone mailing list
- Free World Dialup mailing list
- Two (!) CU-SeeMe mailing lists
- MIT Internet telephony discussion mailing list
- Internet Phone FAQ
- Using the Internet as a telephone FAQ
- Videoconferencing FAQ
- Internet Telephony Interoperability Consortium page

You are, we would imagine, no doubt familiar with the concept of FAQs and mailing lists. Just in case, however, we'd like to refresh your memory:

- A *FAQ*, or Frequently Asked Questions listing, is a compendium (yep — we broke open the thesaurus) of frequently asked questions about a certain subject. Usually moderated by an expert in the subject (though if you read enough FAQs, you may begin to doubt that), FAQs answer the basic questions that everyone has when they are getting started with that particular subject.
- A *mailing list* (also called a *listserv*) is an automated e-mail forum where people can ask questions, answer questions, share tips, and announce new stuff — all to an interested and knowledgeable audience of people who share the same interests. An often-asked question on mailing lists is for configuration advice for particular hardware and software combinations. So if you are having problems, a posting to an e-mail mailing list sometimes jars a response faster than a call to tech support.

Mailing list tricks of the trade

Mailing lists are great sources of information — they can really provide you with a source of in-depth, topic-specific information you just can't find anywhere else. They also can be a big pain in the rear if you don't know what you're doing or if some people who don't play nice are on them. Here are some tips to make your mailing list experience more fun and rewarding:

- Read the list for a while before posting to it. That way you can get a feel for what's going on and not look stupid.
- Not looking stupid helps you avoid getting flamed. *Flaming* is the process of making someone who looked stupid (because they didn't know what they were doing) actually feel stupid by telling them that they are indeed stupid. No one really likes a flame war — at least not any of our friends — so try to stay out of them if at all possible.
- Save the first message you get back from the mailing list after you subscribe. This message typically contains instructions for getting off the list, obtaining back issues, and other neat stuff. If you trash this message and then post to the list asking how to get off, people get mad (refer back to flaming).
- If you get more mail than you can handle, think about subscribing to the digest version of the list. Digests typically compile a day's or a week's worth of messages and send them to you in one big chunk. This chunk can be a great deal easier to digest. Instructions for subscribing to a digest version of a list are typically found in that all-important first message you get when you sign on.
- Moderated mailing lists have an actual human being who screens incoming messages and posts only those that are germane and appropriate. While this technique can potentially reduce the freedom of expression on the list, most moderators are conscientious individuals who are only trying to make the list more useful.
- When you go out of town for a week or two, see if you can temporarily unsubscribe from the list — or get your mail held back for a while. Don't set up your mail program to send automatic replies that say, "I'm out of town for a week; I'll get back to you soon," and then send this message to everybody on the mailing list 30 times. Again, see the flaming section to discover the results of this approach.

By the way, we don't mean to put you off with all this talk of flaming. Our experience on Internet telephony and video mailing lists has been positive, with very few exceptions. We just want to make sure that yours is as well.

VON (Voice on the 'Net) Mailing List

This list, moderated by Jeff Pulver (he's an *Internet World* columnist, among other things, and a name you often see in the Internet telephony world), discusses all Internet telephony products and issues and is probably the biggest mailing list on this subject, bar none. To subscribe, send an e-mail message to majordomo@pulver.com with the subject blank and the words **subscribe von** *first name last name,* substituting, of course, your actual name for *first name* and *last name*.

This list generates a whole bunch of traffic — you may find yourself getting ten or more messages a day from it. After you've been on the list for a while, you may want to switch to the digest version — it can keep your e-mail program from beeping at you all day long.

Internet Phone Mailing List

Another one (of three mailing lists) moderated by Jeff Pulver, the Internet Phone list concentrates solely on VocalTec's Internet Phone product. The process for subscribing is the same as the VON list, except you substitute **iphone** for **VON** in the message body — that is, **subscribe iphone** *first name last name.*

Free World Dialup Mailing List

The last of Jeff's mailing lists (we don't know how he finds the time) is dedicated to the Free World Dialup project — an all-volunteer project being organized to allow noncommercial users to make free gateway Internet phone calls. To subscribe, send an e-mail message (again) to majordomo@pulver.com with **subscribe free world dialup** *first name last name* in the message body.

White Pine CU-SeeMe Mailing List

White Pine Software runs a mailing list for users of its CU-SeeMe program. To subscribe to this list, send a blank e-mail to cuseeme@wpine.com.

Cornell CU-SeeMe Mailing List

Users of the freeware Cornell version of CU-SeeMe aren't left out in the cold either. Cornell runs a mailing list for discussions of CU-SeeMe and announcements of upcoming events. To join, send an e-mail to listserv@cornell.edu with the subject blank and **subscribe cu-seeme-l** *first name last name* in the body. (Again, use your actual first and last name in place of *first name* and *last name.*)

MIT Internet Telephony Discussion Mailing List

MIT, which in the spring of 1996 launched an Internet telephony interoperability project, has also launched a couple of lists on the subject. To subscribe to the Internet Telephony Discussion List, send an e-mail to majordomo@rpcp.mit.edu with **subscribe itel1** in the subject line.

Internet Phone FAQ

Yep, we're back to Jeff Pulver again. He maintains a FAQ on VocalTec's Internet Phone that can be pretty useful for users of that program. Unfortunately, when we last checked, it was a little out of date. But he is updating it as we write, so by the time you go to ftp://pulver.com/pub/iphone/iphone.faq and look at this FAQ, it should be revised.

How Can I Use the Internet as a Telephone FAQ

This FAQ is administered jointly by Kevin Savetz and Andrew Sears of MIT. It covers some basics about how Internet telephony works and provides a listing of products. It's actually posted all over the world, but you may as well go right to its home at http://redwood.northcoast.com/~savetz/voice-faq.html.

Internet Telephony Interoperability Consortium

This page, which is maintained by Andrew Sears, the co-author of the *How Can I Use the Internet as a Telephone FAQ,* focuses on the standards, interoperability, and public policy issues brought up by the advent of Internet telephony. You can find this information on the Web at http://itel.mit.edu, along with a great deal of other information on Internet telephony.

Chapter 22

Ten Good Publications to Read to Follow This Topic

In This Chapter

- Get Techie — weekly and monthly publications
- Newspapers — they're not dead yet
- Makes sense — *PC Magazine*
- Invest in 'Net telephony and video (why not?)

We love to get information online; it's a great way to keep up with late-breaking news, and the hypertext nature of the Web makes going from one topic to another easy. We haven't given up on print media, though — otherwise we'd be pretty silly to be writing a book.

To be honest, until that day when we all have wall-sized, paper-thin, high-resolution monitors that we can use to view Web content (or anything else), paper has some distinct advantages. It doesn't go away when the power turns off, it's just plain easier to read sometimes, and you can carry it with you (some folks like to take it to the bathroom — not us, though).

So even though we are computer junkies, our offices and houses are packed to the gills with books, magazines, and newspapers (just ask our housemates — uh, on second thought, don't remind them).

With that thought in mind, here are some publications we like to read to keep up on telephony, video, and the state of the 'Net in general.

Cabled?

We love *Wired* magazine — no two ways about it. Unfortunately, a great number of people who have never read it think that it's just another 'puter magazine for techno-heads and the like. They're wrong; *Wired* has some techno stuff in it (see the Geek Page), but it's really about the new way of doing things in the digital age. Although *Wired* doesn't track the issue of 'Net telephony all the time, it does cover 'Net telephony, and it also covers the people and issues that this new use of the 'Net involves. We think *Wired* should be mandatory reading. You can find the contents of *Wired* online at `http://www.wired.com`. For additional online-only content, you can also check out *Wired*'s online sister pub, *HotWired* (`http://www.hotwired.com`).

An Old Media Way of Tracking New Media

New Media is a magazine that extensively covers all of the new technologies that will eventually make magazines obsolete. Every issue includes reviews and news about audio and video products, hardware, and other 'Net technologies. Some of the subject matter can be a little techie, but in general it's a good read that can keep you up to date. *New Media* has a Web site at `http://www.hyperstand.com`.

Weeklies for Geek(lie)s

If you are not really a dummy (c'mon, you should have confessed to that earlier in the book — but we're glad you stuck around this far), you may already be familiar with some of the professional weekly magazines that float around the offices and homes of people in the Internet/telecommunications/media (we like the term coined by *USA Today*'s Kevin Maney: *MegaMedia*) industry. Although generally aimed at a more technical market, these publications contain a wealth of information and give it to you in a much more timely fashion than monthly publications are able to (both are slower than online sources, of course).

You may not want to subscribe to these — they're either expensive or available only to a select professional audience — but if you can wander by the room where all of your office's computer folk hang out, you should take the opportunity to look a few of them over.

Inter@ctive Week

Aimed toward the 'Net professionals market, *Inter@ctive Week* contains a whole wealth of information on the Internet, Internet infrastructure, and online commerce. You could get bogged down in some of the reviews of Web servers and wireless data infrastructures (we personally love that stuff, but that's what we get paid to love); skim through that and you can find the latest and greatest on new 'Net technologies. For more about *Inter@ctive Week*, check out its Web site at http://www.interactive-week.com.

MacWeek and *PCWeek*

MacWeek and *PCWeek* are focused on Mac and PC system-administrator audiences, with many columns and opinions by leaders in those fields. They are also sources of up-to-the-minute news on new product announcements and reviews. If you don't have access to these publications, you can view most of their contents online at the Ziff-Davis Web site. (Ziff-Davis is publisher of a whole bunch of different computer magazines, including *MacUser, PC Magazine,* and *PC Computing* — all of which you can check out online.) The Z-D Web site is located at http://www.zdnet.com. Other weeklies you may wish to peruse include *InfoWorld* and *ComputerWorld* — published by none other than IDG! You can find their Web sites at http://www.infoworld.com and http://www.computerworld.com, respectively.

More than You Can Chew?

BYTE magazine is probably the premiere computer technology monthly in the world. It covers all platforms from Mac to Windows to UNIX, and it has extensive coverage of Internet-related issues — including 'Net telephony and video. The only problem with *BYTE* is that it is definitely not beginner-friendly. You can dig pretty deeply into some technical stuff here, and some of the assumptions about the reader's level of knowledge may be a bit optimistic for most of us (us, too — so don't feel bad). *BYTE*'s Web site is located at http://www.byte.com.

Smudged Fingers

Don't overlook what is an increasingly good source of 'Net coverage — your local newspaper. The local papers where we live (the *San Diego Union Tribune,* the *New Jersey Star Ledger,* and the *Boston Globe*), for example, each offer an excellent weekly computer and Internet section that covers the very issues we have been discussing. And even if you aren't lucky enough to live in southern California, Boston or New Jersey, you can probably find the same coverage in your own paper. National papers, such as *USA Today* and *The New York Times,* also offer an increasingly large amount of Internet coverage, so if your local paper isn't plugged in yet, you can still have access to some useful, timely information.

The 'Net Is a World

Internet World, to be precise. Published by Mecklermedia (the same folks who provide "The List" of ISPs we tell you about in Chapter 3), *Internet World* is one of those user-friendly 'Net magazines — quite unlike the ones we mention in some of the previous sections of this chapter. That's not to say that *Internet World* is not informative or interesting, because it is on both counts. In *Internet World,* you can find a great deal of coverage of the latest and greatest on the 'Net; by definition, that means you can find information and reviews about Internet telephony and video products. Check out *Internet World* online at `http://www.internetworld.com`.

PC Magazine for PC Stuff

Another great source for all things having to do with your PC, *PC Magazine* has been devoting a ton of pages lately to Internet telephony and video issues. Like *BYTE, PC Magazine* can get a little bit techie at times, but you can slug through it without too much trouble. You can check out the online version of *PC Magazine* at `http://www.pcmag.com`.

General-Interest Computer Magazines

Don't forget to look in the other general-interest computer magazines — publications such as *Family PC, Mac Home Journal,* and *Computer Life,* to name just a few. They too have jumped on board the Internet bandwagon. So keep your eyes on them; they're sure to be covering the big events in 'Net telephony and video.

Put Your Money Where Your Data Is

One particularly neat magazine, if you are interested in investing some of the money you save by making all of your long distance calls over the 'Net, is called the *Red Herring. Red Herring* covers Silicon Valley firms and other high-tech companies from an investment viewpoint, and some of the companies doing 'Net telephony and video are among the hot companies being watched. Check out the *Red Herring* Web page (`http://www.herring.com`) to look at back issues and for subscription information.

Chapter 23

Help File: Ten Common Problems and How to Fix Them

In This Chapter

- Can't connect to server
- No outgoing sound
- No incoming sound
- Feedback
- Can't adjust the volume
- Poor sound quality
- No video
- Poor video quality
- The mystery of 'Net addresses
- Firewalls — Argggghhhhh!

Internet telephony and video programs are, as a rule, darn easy to use. After you go through the setup procedure and spend a little time adjusting your sound and video (use the tips in Chapter 20 to tweak these), you should be able to start communicating with very little trouble.

A few things, however, may pop up and keep you from having successful connections (or even from connecting at all). Many different programs that you can (and should) try out are available — and the sheer number prevents us from giving specific instructions on every program. We're not going to leave you high and dry, though. This chapter contains some tips that apply to just about every program you may use.

Can't Connect

You can fail to connect in two different ways:

- You aren't properly connected to the Internet.
- You can't get onto the server system of the program that you are using.

If you're not properly connected to the Internet, you probably aren't connecting properly to your ISP. The easiest way to see if you are really connected to the 'Net is to use a Ping program. *Ping programs* basically send out a query to another machine on the Internet (you specify which one) and say, "Hey, are you there and alive?" We generally use the one that is included with Windows, although some TCP/IP software packages come with their own Ping programs. If you can successfully ping someone, you know that you are on the 'Net — so you eliminate that as a problem.

Here's how you ping, using the Microsoft Ping program:

1. **Choose Start⇨Run (or run the Ping program from the File Manager in Windows 3.11).**

   Windows displays the Run dialog box.

2. **Enter the location of the PING.EXE file on your hard drive (probably in the c:\windows folder) followed by a space followed by the IP address or host name of the machine that you want to ping.**

   Choose an address of a computer that should be working — such as `is.internic.net`. In this case, you type **c:\windows\ping is.internic.net** (see Figure 23-1).

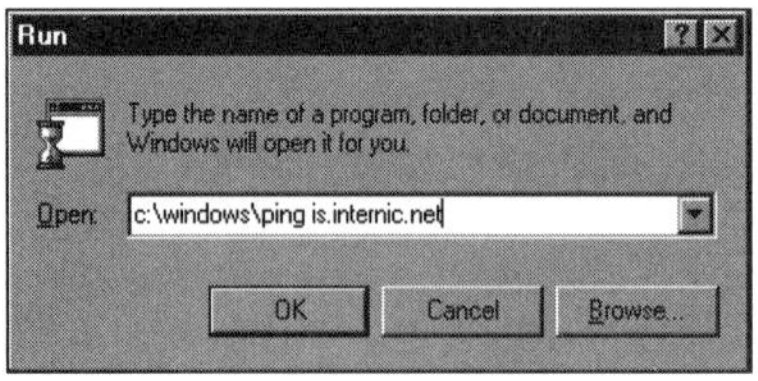

Run

Type the name of a program, folder, or document, and Windows will open it for you.

Open: c:\windows\ping is.internic.net

OK | Cancel | Browse...

Figure 23-1: Pinging InterNic.

3. **Click on the OK button.**

4. **Ping sends some packets out to the other computer, and the other computer should bounce them back to you.**

   You get a screen like the one in Figure 23-2, showing that the other computer has replied.

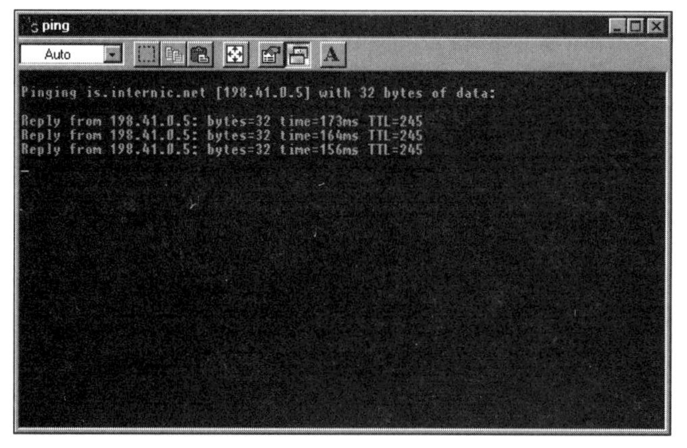

Figure 23-2: Ping results.

If your ping, and pings to a few other known sites (such as your Internet Service Provider or even us — 199.186.159.51 — if you're in a pinch) fail, then you know that you aren't really on the 'Net. Hang up, reconnect to your ISP, and try again. If you still can't connect, it's time to call tech support. Remember, we told you to find an ISP with 24-hour support!

If your ping is successful and you still can't connect to your program's server, chances are that the fault isn't yours; the server may be busy or down altogether. Try one of the following tips:

- If the program has more than one server (like Internet Phone), try to connect to an alternate server.
- If your program allows, try to make a direct connection to someone. Some programs even enable you to connect to yourself for testing purposes.

Unfortunately, one of the downsides of the server model is that you can (and will) have problems with the server on occasion, so there may be nothing that you can do.

No Sound Going Out

If you are connected to someone else on the program you are using but you can't get any audio going out, you should check a few things:

- Is the microphone plugged in?
- Is the proper sound card selected in the configuration or setup window?
- Is another program using the sound card already?

- Do you have the microphone turned on in your mixer program?
- Mac users, do you have the microphone selected as your Sound In source in the Sound Control Panel?
- Is the microphone level turned up?

No Sound Coming In

Most of the time, no sound coming in on your end is a reflection of the previous problem (no sound going out) on the other party's end. If, however, you see signs of incoming sound (that is, the program's display lights up and indicates that the other person is talking) and you still can't hear anything, you can check a few things.

Basically, follow the same advice we give you for sound input problems — only this time, check the output end of things:

- Are your speakers or headphones plugged into the computer — and do they have the power turned on?
- Is the proper sound card device selected in your program's configuration file?
- Is another audio program, such as RealAudio, running in the background?
- Is your sound card mixer set up properly to play back sound?
- Do you have the speaker volume turned up?

Feedback

If you listen to much Led Zeppelin, you already know what feedback sounds like. But if you are not a "metalhead" (and if you don't know what that means, you aren't one), think of the annoying high-pitched sound that people who don't know how to set up a PA system get when they turn on the microphone. Feedback is caused when the output of your speakers is picked up by your microphone and fed back to itself (that is, the sound loops between speakers and microphone until it reaches an annoying, maybe even head-shattering, crescendo).

If you are getting a high-pitched, whining sound from your speakers, feedback is probably occurring. Here's what you can do to alleviate it:

- Move your microphone and speakers farther apart.
- Increase the voice-activation level in your program so that your speakers' output cannot turn the microphone on.

- Turn off the automatic voice activation altogether and toggle the microphone on and off manually.
- As a last resort, switch back to half duplex.

Can't Adjust Your Speaker or Microphone

Most common sound cards (like the SoundBlaster that we use on our PCs) work well with all of the Internet telephony and video programs. However, a few cards out there just don't want to play with the speaker and microphone controls; you can move the sliders up and down and nothing ever happens. If you are in this position, you must use the mixer program that came with the sound card to set your volumes. Sure, we know this is kind of a pain, but you have to do it.

Your best bet is to establish a connection with some patient soul (a friend or some nice person you met on the server) and spend a few minutes setting things up. Have your friend talk with you and, while he or she is sending you an audio signal, open your mixer program and adjust the volume to a comfortable level.

Then adjust your microphone level in the same way; speak into the microphone and adjust the slider or control in your mixer until you are transmitting at a proper level. Keep a close eye on any graphical display that your program shows for microphone levels — don't let it get up into the red zone, because over-modulated (too loud) microphones are one of the biggest causes of poor quality.

The Sound Stinks

What's happening when your sound quality is awful? As we've already told you, sometimes the matter is just out of your control (such as extremely high traffic on the part of the Internet that your data is traveling across). Sometimes, though, other factors affect audio quality — factors that you can control with varying degrees of success. Check these tips out:

- If you are not using the highest-quality codec available for your bandwidth, your sound quality won't be the best. Always begin with the highest-quality codec available to you — you can always move to a lower-quality one if necessary. If the codecs in the program you are using aren't explicitly labeled (that is, high quality versus standard quality), you can figure out which one is the highest quality by looking at the Kilobits-per-second (Kbps) rate of the codec. The highest quality codec is typically the one with the highest Kbps rate.

- The opposite of that principle is true as well. If you try to use a codec that requires more bandwidth than you have available, sound quality suffers. Codecs are really something you may just have to play with until you get things optimized.
- Make sure that you're not wasting bandwidth by doing other things in the background. FTPing files, opening graphics-heavy Web pages, or sending huge whiteboard files in the background all take up big hunks of your available bandwidth — resulting in lost packets and bad sound quality.
- Use full duplex only if your computer has enough horsepower to handle it. If you're using a slow 486 (like an older 33 MHz model), you may be able to switch on the full-duplex mode in the program that you're using — but when your CPU tries to compress your outgoing voice and decompress the incoming one at the same time, it will get overworked. Try switching to half duplex and see if that approach helps.
- If you are using a videoconferencing program and your audio is not working well, try freezing your video while you are transmitting audio. This trick gives your audio more bandwidth.

No Video

So everything is running fine, you just talked to some nice person on your audio telephony program, and he said, "Hey, I've got CU-SeeMe. Want to videoconference?" And you're thinking, "Great, I've been meaning to try out the direct person-to-person conferencing feature." So you fire up CU-SeeMe and look over at your local video window and see . . . nothing. What to do?

First, don't panic. This situation happens sometimes. Here are a few things to look at:

- Go into your video source Preferences menu and make sure that the video digitizer you are using (the QuickCam or your video capture card) is actually the one that you selected.

  Don't forget to check the accuracy of your video settings. For example, most programs require black-and-white QuickCams to be set at 64 shades of gray; set it at 16 accidentally, and you receive no video whatsoever.
- If that doesn't fix it, make sure that your camera is actually plugged into the back of your computer and is powered up (never forget the old ON/OFF switch).
- If you're still having trouble, make sure that another program that also uses your video source isn't open. If you were just using FreeVue and forgot to quit the program, for example, FreeVue will still be monopolizing your video digitizer.

- When all else fails, disconnect, shut down your computer, and start over. You'd be surprised how much benefit you can get from a good old-fashioned restart.

Poor Video Quality

As we've mentioned several times already, video quality is usually engaged in a tooth-and-nail battle with audio quality for the limited bandwidth that most 'Net videoconferencing users can access. So if your video quality isn't all that great, you may not be able to do much about it.

First, make sure that you follow the rules (we list them in Chapter 20) for optimizing the quality of your video presentation. If your video still stinks, do the following:

- Go into the program's Preferences or Settings menu, increase the proportion of bandwidth allocated to video, and decrease the amount allocated to audio by an equivalent amount.

  Note: We usually look at video as the less-important part of conferencing — audio is our number-one priority — but if the video is more important to you, this can help.

- Reduce the number of open video windows on your machine. Each video stream you receive takes up a chunk of your total bandwidth, so fewer open windows equal more bandwidth per window.

Demystifying IP Addresses

By now, you are familiar with the concept of dynamically-assigned IP addresses — the kind of addresses that are such a nuisance when you are trying to use the direct-connection method (making a person-to-person call on NetPhone, for example). Unfortunately, sometimes you even bump into a problem with connections to IP addresses you know are correct — addresses you've called several times in the past. What's going on?

- The computer's address on the network has changed. This situation happens more often than you'd think — especially on corporate LANs where people come and go, employees change offices, and new equipment is installed.
- A firewall was installed somewhere between you and the other person (it could be at either end), and it is blocking your call.

- The problem could be as simple as the fact that the other computer isn't running the program you are trying to connect on — you could spend all day trying to dial someone up (using the correct IP address) and fail if no program that can answer your call is running. You'll often find this to be the case for CU-SeeMe reflectors that are only up and running part of the day.

You can take no direct actions to fix this problem unilaterally; you have to pick up the phone (yes, it is a good thing we are still planning to keep these things around), call the person you are trying to contact, and find out what's going on. Of course, you can also send him or her a fax, fire off an e-mail message, drop by, and so on.

Punching a Hole

Got a firewall between your computer and the outside world? Better make friends with your system administrator. Many (probably most) of the 'Net telephony and video programs that we talk about use UDP (the transport protocol we discuss in Chapter 2). That's okay for most people, because UDP connections generally have less delay than TCP connections (the other kind of transport protocol we mention in Chapter 2); but unfortunately, most network firewalls block just about all UDP ports.

Even if your program supports and uses the RTP protocol, it still sends data through TCP and UDP ports on the firewall, and it may still run into firewall problems (Well, we call them problems; the people who sell firewalls will say that the firewall is doing its job — blocking packets of data that it's not sure about!).

What are you to do? Three things:

- If the program gives you a choice of using UDP or TCP (and a few do), choose TCP — you'll most likely have no problems connecting then.
- Check in the Preferences or Settings dialogs of your program — a few, such as Intel Internet Video Phone, have a special dialog box for setting preferences to get around, or through, a firewall. You need to get some information from your system administrator to complete this dialog box. In fact, we recommend that you just ask your system administrator to come over to your computer and fill out the dialog box for you.
- Ask, beg, plead, bribe, or otherwise convince your system administrator to manually "punch a hole" in the firewall for the UDP ports that your program uses. She or he will know what this means (although a call to the program's technical support number may be required). Bring plenty of low-denomination unmarked bills for this one. The good news here is that many firewall vendors are starting to design in special support for H.323 telephony programs, so configuration is getting to be a snap.

Chapter 24

Ten Neat Things You Can't Do on the Net — Yet!

In This Chapter

- Call users of other products
- Call on the 'Net from a regular phone
- Use your television on the 'Net
- Watch high-quality video
- Listen to CD-quality music
- Forget voice mail — leave video mail
- Watch *Seinfeld*
- Do some real virtual shopping

In this chapter, we talk about some things that you can almost do now on the 'Net — things that, if 'Net technology keeps progressing the way it is, you'll be doing before you know it.

Universal Product Interoperability

When we say *product interoperability,* what we really mean is this: Can I use my copy of Internet Phone to talk to my pal who has CU-SeeMe? The answer, unfortunately, is often still *no* (with a few notable exceptions).

The problem used to be one of standards — or, more accurately, lack of standards. As we tell you in Part I of the book, things on the Internet work because of standards. E-mail goes from your Macintosh connected to your Netcom Internet account to your mom's Windows PC on her AT&T WorldNet account because all e-mail programs adhere to established and widely accepted standards. But hey — you may say — doesn't Part I of this book talk about H.323 — an ITU (International Telecommunications Union)-approved standard for Internet telephony? Yes, it does.

So what's the problem? Well, there isn't one really, except time. Only a handful of H.323-compliant programs exist right now, and a huge installed base of popular programs all use their own proprietary means of talking or sharing video. The developers of most of these proprietary programs plan to offer standards-compliant products a version or two down the road — but it takes time to make the necessary changes without abandoning the things that make these programs popular with their users. And then people have to be willing to pay extra to upgrade.

A few minor glitches also hinder the acceptance of the standards — such as some wrangling between various companies over royalties and intellectual property rights for the various codecs used by standards-compliant Internet telephony programs. And many Internet telephony programs still use different, and incompatible, directory service methods — so connecting even two H.323 'Net telephony programs requires the use of a third-party directory, such as Four11. But these stumbling blocks are really pretty minor, we think, and all new standards face these types of issues.

We think that this standards problem has been solved — H.323 is THE standard for Internet telephony. Not a week goes by without someone announcing their support of H.323, and in due time it will be the rare 'Net telephony program that isn't compliant with the standard. Be patient — we'll get there!

Using Your Phone to Make 'Net Calls

As we discuss in Chapter 15, the technology already exists for making 'Net calls via your phone. The gateway products that enable phone-to-phone calling over the Internet are out there — for sale and ready to be installed. People already use these products in corporate intranets. So why are we saying that you can't do this yet? Mainly because chances are good that YOU can't. A lag always occurs between development of such radical new technology and its deployment in a sizable number of markets.

A few companies have already announced that they will soon offer global long distance service over the 'Net. We believe them, and think that within a year or two this kind of service will be widely available. Right now, it's not. You can only call between specific locations or specific cities.

In the meantime, you may end up using Internet telephony from your regular telephone without even knowing it. The deployment of things like MCI's Vault technology and Internet telephony-based call centers means that the customer service rep you connect to from your regular telephone may actually be sitting in front of a PC on the other side of a gateway somewhere. Think about that the next time you dial that 800 number to order some gourmet dog biscuits.

Videoconference with Everybody

Today's 'Net-based videoconferencing programs, such as CU-SeeMe and NetMeeting, can be great tools for business or personal interaction. They are limited, however, because you can connect only to others who are using the exact same program, or others using an H.323 Internet videoconferencing program (if the program you use is H.323 compliant). Many other videoconferencing products that you can't yet access over the Internet are in use throughout the world.

This situation is changing as gateways between H.323 videoconferencing programs and other types of videoconferencing programs are developed. Pretty soon, probably within the next few years, you will be able to participate in a videoconference in which some of the participants are connected by the 'Net, some by ISDN desktop videoconferencing products, and still others by stand-alone videoconferencing equipment (the fancy stuff that looks like something out of *Star Trek*).

High Quality Video

A good quality, television-like video signal, such as the one put out by the new DVD (Digital Video Disk) that is rolling into stores this year, takes up at least 3.5 Mbps (Megabits per second) of bandwidth. Compare that to the 28.8 Kbps modem that connects your machine to the 'Net today, and you can see why making Internet video work is so hard. (Let's see, 3.5 Mbps divided by 28.8 Kbps, carry the 17, hmmmm . . . that equals about 121 — so that signal is about 121 times as large as the pipe that would carry it to your house — see the problem?)

That DVD signal is already severely compressed by using MPEG-2 (which is pretty much at the limits of our technology) to compress a video signal and still make the video signal appear crisp and clear on a TV screen. So we are butting up against the old compression versus bandwidth tradeoff again.

Someday you'll be able to get a picture from the 'Net that's as good as the one that your $199 Sony gives you — but not yet. Sorry. But as the access bandwidths increase (à la cable modems and ADSL, as we discuss in Chapter 25), your increased bandwidth will tremendously increase your reception. (No more rabbit ears to mess with!)

Digital Television, WebTV, and TalkTV

A new wave of television sets is heading for the market — digital television, which promises clearer pictures, better options, . . . and surely higher prices. In another decade, most television sets will probably be digital, just like most TV sets are color today.

Microsoft recently poured money into WebTV, a company dedicated to bringing the Internet to the world's couch potatoes. The key goal of the Microsoft-WebTV relationship is to bring Web browsing capabilities to your TV via a computer chip that fits within the TV set instead of relying on the bulky set-top box that exists today.

When televisions start shipping with Web browser software and everything that Microsoft can bring to bear for the consumer Internet marketplace, you can expect that Internet telephony will be part of the action. So before long, you may be talking to your TV set instead of your computer. You'll enjoy all of the same benefits of Internet telephony and video as you do now — only on a bigger screen and in the living room.

Of course, this evolution begs the question: When will the television set turn into a computer?

CD-Like Sound

Sound quality faces the same problems on the Internet as video does, but fortunately to a much lesser extent because the size of an audio signal (even the one coming uncompressed straight out of the back of a CD player) is much smaller than that of a video signal.

That said, the most common means of delivering live audio over the 'Net today (such as RealAudio) can sound good, but these methods are not quite up to the standards of FM stereo radio yet — not to mention the sound quality of a CD.

Prospects for higher-quality music delivery over the 'Net (and we are talking about music because such things as stereo and frequency range don't matter as much for voice) are good. In fact, StreamWorks already offers stereo 44.1 kHz streams of several music sources — 44.1 kHz is the same sampling rate that your CD player offers — if you have an ISDN or higher connection.

Why Not Say It with Pictures?

As we write this, several of the major Internet telephony programs either already offer, or will soon offer, a voice-mail system integrated into the program — a system that enables you to leave a voice-mail message for someone who is not online or is busy when you call. The next logical step for this kind of system is to integrate it into a videoconferencing program so that the program enables you to leave a video-mail message for someone when you can't connect.

Sounds like something out of science fiction, but there's really no reason why it can't be done very soon. We can't wait.

Hey, Kramer

Video on the 'Net, as it stands now, is most useful for viewing small clips on demand — great for showing a quick news bite or showcasing a product — but video on the 'Net is not really designed for broadcasting to millions (or even just thousands) of people at a time. Conventional television systems are best at feeding the same signal to an immense number of people at once.

A system on the 'Net called MBONE (Multicast Backbone), though, can perform a broadcasting function efficiently and effectively. The MBONE is really just a network of specialized switches and routers within the greater network that makes up the Internet. The MBONE makes use of a principle called *multicasting.* Multicasting sends video (or audio or text or anything, really) data out over the 'Net only once, and the data reaches everyone in its target audience without ever passing over the same sections of the network twice.

The majority of other 'Net video streaming and videoconferencing products use a technique called *unicasting,* which means that each and every recipient of the video data gets his or her own unique transmission of the data. As you can imagine, this technique adds greatly to the complexity and the bandwidth requirements when you start talking about trying to broadcast *Seinfeld* to its millions of weekly viewers.

The MBONE has the following drawbacks:

- Only a small part of the Internet is part of the MBONE, and chances are good that your ISP isn't on it.
- The MBONE was designed for high-bandwidth dedicated access — like the dedicated multi-megabit connections common among the universities, corporations, and research centers that are connected to it today.

Even if you were somehow connected to the MBONE by your ISP, your 'Net connection (even ISDN) would probably be too small to allow you to hook up to it.

So it will probably be a while before the 'Net replaces your television for watching mass broadcasts of video programming. That's okay, though. Use your television for what it is best for (viewing broadcasts) and use the 'Net to view the kind of specialized, on-demand video that it is best at providing.

The good news here is that a consortium of Internet video vendors has formed to support and promote the widespread development of multicasting on the 'Net. With the kind of horsepower found in this organization (Microsoft, for example, is one of the founding members), we wouldn't be surprised to see multicast a lot sooner than anyone expects.

If you are lucky enough to be part of the MBONE (or if you want to find out what you need to do to get involved in it) and you want to get smart on the subject, read *MBONE: Multicasting Tomorrow's Internet,* by Kevin Savetz, Neil Randall, and Yves Lepage (published by IDG Books Worldwide, Inc.).

Full-Service 'Net Shopping

This one isn't too far away from realization. Pretty soon, by using existing technologies, you'll be able to point your Web browser at a site, listen to audio descriptions of products, view videos of them in action (maybe even in 3-D), search through databases to find the options you want, and then talk (or even videoconference) with a sales representative. When you decide what you want to buy, you can even place your credit card order by using a secure transaction feature in your browser.

The Web-based call centers that we talk about in Chapter 14 are an example of this kind of 'Net shopping scenario — and they are drawing a lot of interest from Web retailers.

And you thought buying by catalog was cool.

Three to Beam Up

Unfortunately, no matter how cool 'Net telephony and video technologies become, we are pretty confident that you will still have to wait for the courier to come by and drop off your packages for you. Particle beams and transporters are just going to have to wait a bit longer. *Star Trek* still is a *bit* futuristic.

Chapter 25

Ten Things About to Happen with Internet Telephony and Video

In This Chapter

- Working together
- Integration migration
- MMX horsepower
- More speed — ISDN
- Even more speed — cable modems
- Yep, more speed again — ADSL
- Gateways galore!
- The 'Net gets faster and friendlier
- ISPs get dirt cheap
- Quality, at a price

In this chapter, we look into our crystal balls and tell you some of the ways that Internet telephony will soon get better and easier. Actually, we didn't have to do that much reading of tea leaves. Much of this stuff has already been announced, and many smart people are putting time (and money) into making these technologies and initiatives into real-world products that everyone can use soon. We're not talking about five or ten years from now either; you'll see most, if not all, appearing in the next one to two years.

Interoperability Becomes Reality

Time and time again throughout this book, we bring up the subject of H.323 and other standards. We aren't just doing this for the fun of it — standards, and their acceptance, are vital to the continued success and growth of Internet telephony. The good news is that we've got a robust set of standards (H.323, T.120, LDAP, and so on) that work and good products available (NetMeeting, Conference, and Intel Internet Video Phone) that use them.

The bad news? Well, lots of work still must be done. Many of the most widely-used programs don't yet adhere to the standards — and not all of those that do are completely compliant. And the standards themselves aren't set in stone. In fact, a great deal of controversy exists in the industry right now about licensing and royalties for the default codecs used in H.323.

The industry will continue to experience some growing pains as standards-based Internet telephony becomes more widespread. But don't worry, because it will happen, and soon. None of the vendors of Internet telephony products — software or hardware — can afford to buck this trend.

Phone? Web Browser? What's the Difference?

If you read the Internet industry trade press, you'd certainly be forgiven for assuming that Netscape and Microsoft are two companies that just can't see eye to eye. A battle always seems to be brewing — Java versus ActiveX, Microsoft cascading stylesheets versus Netscape cascading stylesheets, Channel Definition Format versus something or other. (If this sounds like technobabble to you, don't worry; just realize that each company is trying to outmaneuver the other in the battle for the hearts and minds of Web users.)

Luckily for us, things usually work out, and we don't end up with two completely incompatible Web browsers. In the Internet telephony field, surprisingly, these two giants appear to be heading in roughly the same direction.

Both companies have released H.323-compatible products as part of their most recent Web browser packages. (Microsoft did it first, and Netscape followed shortly behind.) And both companies are pursuing a high degree of integration between their phones and the rest of the browser package. Interfaces are similar, and address books are shared, for example.

We think this is the beginning of a trend. As the browser becomes a more and more integral part of the computer interface (almost merging with the desktop), the Internet telephony features will become less and less part of a stand-alone program and more like a module within the browser software.

Additionally, you'll find Internet telephony software being integrated into other Internet programs. Already you can find voice capabilities built into things like online games (so you can yell "gotcha you #$@&% !" as you finish off your opponent) and into group browsing and chat programs like PowWow. We think that in the not-so-distant future 'Net telephony will move away from the stand-alone programs that we see today and will become more and more tightly integrated into other 'Net programs.

The Sign Says Merge

In Part III of this book, we tell you a bit about gateways and how they allow Internet telephony and traditional telephony to coexist — and work together — peacefully. Until now, the most common use of these products has been within a corporate intranet or within the network of a specialized service provider offering a low-cost service like international long distance over the 'Net.

We're here to tell you that this is just the beginning. As Internet telephony and gateway technology become more mature, you'll find that the line between the old and the new becomes increasingly blurred. In fact, the goal of gateway producers and service providers is to make the whole Internet telephony experience completely invisible to you, the user.

Browsing a Web site and click on a button to place an order? The sales rep you talk to may also be on an Internet phone or could be on the PSTN network. You'll neither know nor, if things are running correctly, care.

The opposite case could be true as well. Call a company's 800 number from your regular telephone, and you may connect to a call center in Durham, North Carolina. But the service technician you get patched through to could very well be in Portland, Oregon, sitting in front of and talking into a PC. The gateways will be seamless and invisible.

Speed Demons

We spend a considerable amount of time earlier in the book discussing the role of bandwidth and compression in determining the quality of Internet voice and video communications. For most people, the amount of accessible bandwidth is very small (28.8 Kbps modems in most cases), so programs that use the Internet for voice and video communications have to compress the digital signals they send out so that these signals fit the pipes running into our houses.

The problem comes when you try to do a really large compression — such as trying to squeeze a huge video signal over your modem. This situation taxes your computer's capability to perform the compression and decompression fast enough to provide you with decent audio and video signals. Remember that the complexity and size of the calculations that your computer must perform are enormous; with such large compressions, your computer's CPU can reach a point where it just can't keep up.

You can view the whole process of compressing audio and video as a three-sided equation. You have quality on one hand, bandwidth compression on the other, and computational complexity on the, ahem, third hand. What you want is high quality, high compression, and low complexity. Unfortunately, you can't have it all — something's got to give. Traditionally, you've sacrificed either quality or bandwidth because PCs just haven't had the CPU horsepower to handle the complex computations required for high quality and low bandwidth at the same time.

That's where new high-powered CPUs such as the Intel MMX come in. With some special multimedia-processing functions built right into the MMX Pentium chip itself, this chip can perform video and audio compression functions a great deal faster than a regular Pentium chip can. The MMX chip became widely available in regular consumer PCs in early 1997. Mac users aren't being left behind either — the most powerful desktop PCs available as we write have PowerPC chips under the hood, and Apple is developing a new chipset called TriMedia that will do the same sort of special processing that MMX performs.

'Net telephony and video programs that are written to take advantage of MMX (and they will have to be rewritten to optimize the use of the technology) or fast PowerPC chips will fit an even bigger, higher-quality signal over the same size pipe, so modem users can get even more of the signal across the 'Net. We can't wait.

ISDN for Everyone

ISDN isn't particularly new, and it isn't particularly revolutionary at this point in time, nor is it the fastest technology around. Indeed, new access technologies are coming that are orders of magnitude faster. But ISDN is available in the here and now, and it can really improve the quality of your 'Net communications. It's nearly five times faster than the fastest modems out there right now.

Because ISDN is so readily available, people are clamoring for it (and telephone companies and ISPs are starting to listen). Unless some even faster technologies (which we talk about next) come up and steamroller ISDN very soon, you can expect to see ISPs and telcos offering reasonably priced, all-in-one ISDN/Internet access/Internet hardware packages that can get you online in a digital way, quickly and easily.

If one such deal becomes available in your area, take a close look at it; it could be just what you need to boost the quality and reduce the frustration of using 'Net video and audio products.

For more information on ISDN, check out Dan Kegel's ISDN page at `http://www.alumni.caltech.edu/~dank/isdn/`.

HBO and the Internet, Too!

Of course if you think that ISDN is fast (128 Kbps), wait until you see this one — cable modems. No doubt you've heard them mentioned; they've had 'Netizens drooling for a couple of years now.

Basically, a *cable modem* uses the existing coaxial cable that plugs into the back of your TV to carry very-high-speed data into your computer; the cable companies promise up to 10 Mbps.

If carrying data on cable-TV cables is so easy, why don't we all have cable modems? The reason has to do with the architecture of the cable-TV network: The cable-TV network was designed to go in only one direction, with the signal flowing outward from a central point (the cable company's local office) to a bunch of end points (everyone's TVs). And unlike the traditional phone network, a cable-TV network has no switches that can direct those little electrons back and forth between different points on the network.

Some of the bigger, more aggressive cable companies are working feverishly at installing the network equipment that will make cable modems work, and big companies like Motorola are rolling the plants at full speed to build the modems themselves.

It won't be long before your local cable company offers you high-speed 'Net access. If everything works as planned, it will be so fast that it will make your eyes water.

And that's not the best of it — you'll have a full-time connection into the Internet. This means that your computer will be available to take calls 24 hours a day (subject to cable company reliability, of course, which leaves a great deal to be desired). So you will always have an IP address, you won't have to worry about setting prearranged times to meet people on the 'Net for phone calls, and you can start treating your computer like your regular telephone. Ain't that sweet!

One thing to look out for: In their zeal to get something out there fast, some cable companies are putting in cable modems with a low-speed, 28.8 Kbps return path. In other words, 10 Mbps to your home, 28.8 Kbps back. This is not the greatest for symmetrical applications like videoconferencing, in which you get as much information as you give. Watch out for this low-bandwidth return path; it can limit what you can do with your connection.

If you want to find out even more about cable modems, take a gander at Sam's Interactive Cable Modem page at `http://www.teleport.com/~samc/cable5.html`.

ADSL: Like Fiber, but It's Copper

When it comes to offering high-speed Internet access, your local telephone company has a problem that is the exact opposite of the one facing your cable company. The telcos have, collectively, hundreds of billions of dollars invested in super-fancy switching devices, fiber optic cables, and local offices in every town, but running from those central offices to everyone's home is a skinny little pair of copper wires.

Until recently, no one could figure out how to get much more than 28.8 Kbps modem or 128 Kbps ISDN speeds over these lines — not much of an on-ramp to the Information Superhighway of the future. (There — we used the term "Information Superhighway" in our book — this is the last chapter, and we knew that we couldn't help but slip it in somewhere.)

Replacing those copper wires with something else, like fiber optic cable or even the coaxial cable that cable TV uses, will remove this bottleneck. Someday your phone company will make this change. For now, though, telephone companies have come up with a new kind of modem technology called ADSL (which stands for Asymmetrical Digital Subscriber Line, if you must know).

ADSL (and related technologies that replace the *A* with an *H, S,* or *V*) allow you to get multi-megabit speeds (somewhere in the 1.5 to 6.0 megabits-per-second range) over existing phone lines. This speed is more than competitive with the speed offered by cable modems.

So telcos can boost your access speed 100 times or more without spending the serious bucks required to run new kinds of lines to everyone's house. And the telephone companies are talking about offering you this for $30 to $50 per month. Get outta here!

You'll see some telephone companies offering ADSL-based Internet services to their customers very soon — probably by the end of 1997. If you want to read a whole bunch more information about ADSL technology and news, check out our own Web page on the subject at `http://www.xdsl.com`.

Bigger, Stronger Backbones

The other bottleneck that your real-time audio and video signals face is the backbone of the Internet itself. For the purposes originally envisioned by the designers of the 'Net, delays and slowdowns caused by high traffic and the transport protocols used on the Internet may have been a slight annoyance, but nothing more. Who really cares if it takes 30 minutes instead of 10 for an e-mail to get from one place to another (or if it takes an extra minute to finish downloading a file that you are FTPing from another computer somewhere)? Sure, these situations can be a little frustrating, but in the end, you get what you came for.

Real-time applications such as Internet telephony, on the other hand, can be seriously degraded, or even fail to work at all, because of these excessive delays or lack of bandwidth. Luckily, things are getting better.

First of all, the companies that provide most of the backbone bandwidth of the Internet (by this we mean the big ISPs and network providers like MCI, UUnet, BBN Planet, and the like) are rapidly upgrading the capacity of their fiber optic networks. MCI, for example, was in the middle of installing new switching and router technologies that would triple its backbone capacity from 55 to 155 Mbps while we were writing this (as were many other companies).

Secondly, new protocols (such as RTP and RSVP) are being adopted as Internet standards. These protocols, and others like them, are better suited to the real-time nature of 'Net multimedia communications and to the bandwidth requirements that features like video have.

Cheap, Cheap 'Net Access

Low-cost Internet access is already here. In the past year we've seen companies such as AT&T and MCI offer low-priced, unlimited Internet access services, and we've also seen many of the established ISPs come out and meet their prices. For example, when we wrote the first edition of this book last year, the Netcom dial-up service that we used for testing products over a 28.8 Kbps modem connection dropped in price by more than half in about two months. Since then, just about every major ISP has dropped its prices similarly, and some startup companies are trying to offer even cheaper access than the industry standard of $20 per month.

This price-reduction phase will continue, as more and more companies compete for your Internet access business. We wouldn't be surprised to see phone companies and others throwing in free Internet access as part of a package of other services. So getting online could become cheaper and easier than ever.

Pay to Play

Now we're going to totally contradict our previous statement and say this: Pretty soon you'll see Internet access that costs more! There, we got it out. Sounds like heresy, doesn't it? Well, despite the continuing trend for Internet access to get cheaper over time, a small, but growing, counter-trend has developed for more expensive access. ISPs are building up their networks, and network equipment providers such as Cisco are developing systems that can provide users with Quality of Service guarantees, all of which means your ISP will soon be able to offer you higher quality service — for a price!

We think that in the near future you'll see ISPs offering different levels, or tiers, of Internet service. At the bottom tier will be the cheap, unlimited access to which we've all become accustomed. At this price, you'll get no guarantees, just a modem connection and an IP address. At the other extreme, you'll see some sort of pay-as-you-go service that gives your data priority as it flows across the network and guarantees adequate bandwidth for what you're doing. In between will be . . . , well, in-between options. How much will it all cost? Sorry, our crystal balls aren't that clear.

Part V
Appendixes

In this part . . .

Just because they're in the back of the book doesn't mean you should ignore the appendixes.

The first appendix, Appendix A, helps you track down an Internet Service Provider so that you can get online.

Appendix B gives you some more information about the Internet telephony gateway products that we discuss in Part III — in case you want to explore installing one in your business.

Appendix C helps you load the CD-ROM that accompanies this book, install programs, and start using Internet telephony and video programs RIGHT NOW! Why wait for downloads to finish? We wouldn't.

Appendix A

U.S. National Internet Service Providers

This appendix contains a far-from-complete listing of the big U.S. ISPs. You can find literally hundreds of other, smaller providers, many of which specialize in specific geographic locations. Don't overlook these smaller providers, because they may have a great deal to offer — especially in terms of service and the availability of special features, such as ISDN access.

If you already have some sort of Web access (maybe through AOL or CompuServe, or at work, or on a friend's computer), you can use that access to check out some of the online sources that provide helpful information for choosing an ISP. Our favorites include the following:

- Mecklermedia's The List, at `http://www.thelist.com`.
- The Yahoo! listing of ISPs, at `http://www.yahoo.com/Business_and_Economy/Companies/Internet_Service/Internet_Access_Providers`.

  If you're searching for an ISP outside of the United States, the Yahoo! site is especially helpful.
- c|net, at `http://www.cnet.com`. c|net recently conducted a member survey, with thousands of c|net subscribers rating their own ISPs.

This list is sure to grow soon, as some of the major telcos (such as Bell Atlantic and Pacific Bell) and cable companies start to roll out their own Internet service offerings.

Internet Service Providers

Provider	*Web Site*	*Phone Number*	*ISDN*
AOL	`www.aol.com`	800-827-6364	No
AT&T WorldNet	`www.att.com/worldnet`	800-967-5363	No
Concentric Networks	`www.concentric.com`	800-939-4262	No
Earthlink	`www.earthlink.com`	800-395-8425	Yes
Epoch Internet	`www.hlc.com`	800-915-5515	Yes
IBM Internet Connection	`www.ibm.net`	800-455-5056	Yes
MCI Internet	`www.mci.com`	800-550-0927	Yes
Mindspring	`www.mindspring.com`	800-719-4332	Yes
Netcom Online Services	`www.netcom.com`	800-353-6600	Yes
Prodigy Internet	`www.prodigy.com`	800-776-3449	No
Sprint Internet Passport	`www.sprint.com`	800-747-9428	No
SPRYNET	`www.sprynet.com`	800-777-9638	No
Whole Earth Networks	`www.wenet.net`	800-246-6587	Yes
UUNET Technologies	`www.uunet.com`	703-206-5600	Yes

Appendix B

Gateway Hardware Vendors

This appendix contains a listing of vendors of Internet telephony gateway products. This is a fast-growing group — many telecommunications and Internet equipment vendors are beginning to enter this expanding marketplace.

Most of the companies who provide these products are members of the Voice over IP Forum — a group organized by the International Multimedia Teleconferencing Consortium, or IMTC. The IMTC has a complete listing of member companies on its Web site (`http://www.imtc.org`) — this is your best starting point if you are interested in finding new entrants to this market.

Gateway Hardware Providers

Vendor	*Web Site*
Brooktrout Technology	`http://www.brooktrout.com`
Dialogic	`http://www.dialogic.com`
Lucent Technologies	`http:/www.lucent.com`
MICOM Communications	`http://www.micom.com`
Natural Microsystems	`http://www.nmss.com`
NetXchange	`http://www.ntxc.com`
RADVision	`http://www.radvision.com`
Rockwell	`http://www.rockwell.com`
ViaDSP	`http://www.viadsp.com`
Vienna Systems	`http://www.viennasys.com/`
VocalTec	`http://www.vocaltec.com`

Appendix C
About the CD

Here's what you can find on the CD-ROM that accompanies *Internet Telephony For Dummies,* 2nd Edition:

- Full and demo versions of some leading Internet telephony software
- AT&T WorldNet software for Internet access
- Some cool extra software to help get you up and running

For Windows users, we even include a special interface that makes it a lot easier to see what's on the CD, get more information about each program, and install the programs you want to try out. Mac users don't get the special interface because we think the Mac interface is so intuitive and easy to use that you don't need it.

System Requirements

Make sure that your Apple Macintosh or compatible meets the following system requirements for using this CD:

- A computer with a 68030, 68040, or PowerPC processor.
- System software Version 7.1 or higher (System 7.5 is recommended).
- A CD-ROM drive — double-speed (2x) or faster.
- At least 4MB of free RAM for most programs. (*Free RAM* is the amount of memory available to the Mac when no other programs are running. You can check your available memory by clicking the Apple menu in the Finder and choosing About This Macintosh.) In some cases, turning on virtual memory will help you if you are a little short of memory, but some demos may run erratically.
- A monitor capable of displaying at least 256 colors or grayscale.
- A modem with a speed of at least 14,400 bps.
- At least 22MB of hard drive space available to install all the software from this CD. (You'll need less space if you don't install every program.)

PC users should meet the following system requirements for using this CD:

- Windows 3.*x* (that is, 3.1 or 3.11), Windows 95, or Windows NT installed.
- If you're running Windows 3.*x*: a 386sx or faster processor with *at least* 8MB of total RAM.
- If you're running Windows 95 or Windows NT: a 486 or faster processor with *at least* 8MB of total RAM (16MB of RAM recommended).
- At least 31MB of hard drive space available to install all the software from this CD. (You'll need less space if you don't install every program.)
- A CD-ROM drive — double-speed (2x) or faster.
- A sound card with speakers.
- A monitor capable of displaying at least 256 colors or grayscale.
- A modem with a speed of at least 14,400 bps.

If you need more information on PC or Windows basics, check out *PCs For Dummies,* 4th Edition, by Dan Gookin; *Windows 95 For Dummies* by Andy Rathbone; or *Windows 3.11 For Dummies,* 3rd Edition, by Andy Rathbone (all published by IDG Books Worldwide, Inc.).

Using the CD

For Windows Users

Installing the bonus software is easy, thanks to the CD interface. An interface, as far as this CD goes, is a little program that lets you see what is on the CD, gives you some information about the bonus software, and makes it easy to install stuff. It hides all the junk you don't need to know, like directories and installation programs.

If you have Windows 95, follow these simple steps to get to the CD interface

Put your CD in your computer's CD-ROM drive. Click on the Start button, and then click on the Run option in the Start menu. In the Run dialog box, type **D:\setup.exe**. If your CD-ROM drive is not called D, be sure to use the correct letter for your drive. The first time you use the CD, you see a License Agreement. After you agree to the terms of the agreement, the interface opens, and you can start browsing the CD. Follow these steps any time you want to use the CD.

If you have Windows 3.1, follow these simple steps to get to the CD interface

Put your CD in your computer's CD-ROM drive. Click on the File menu header in Program Manager and then click on the Run option in the File menu. In the Run dialog box, type **D:\setup.exe**. If your CD-ROM drive is not called D, be sure to use the correct letter for your drive. The first time you use the CD, you see a License Agreement. After you agree to the terms of the agreement, the interface opens up, and you can start browsing the CD. Follow these steps any time you want to use the CD.

Using the CD Interface

The CD interface has three general regions: the category list, the product list window, and the background. When you first see the interface, the product list window displays instructions on using the CD. The category list on the left side lists the categories of software that are available on this CD. When you select an option from the category list, the product window changes and shows you all of the products available in that category. Now you can select a product from the list and get more information about it or install it.

To get more information, click on the Info button. A window opens up — probably your NotePad program — with a brief description of what the program does and any special information you may need to know about installing it. After you read the information, be sure to close the text window, or you get text window buildup — that is, windows piling up all over your desktop. To close the text window, just click on the File menu and choose Exit. To install a program, just click on the Install button. The programs' installers walk you through the process.

For Mac Users

For Mac users, we decided to take advantage of the simplicity of the Mac Finder, so all you need to do is pop the CD into your CD-ROM drive. When the CD icon appears on your desktop, double-click on it. The CD window opens up. Be sure to read the License Agreement and the Read Me files. These contain important information about the programs on the CD. To install programs to your computer, simply run the program's installer, or copy the program's folder to your hard drive. Installer programs generally have names that end in ".sit" or ".sea" — some are simply called "Program Installer."

What You'll Find

Here's a summary of the software on this CD.

AT&T WorldNet Service

In the Cool Stuff category for Windows users; in the Cool Stuff folder for Mac users. For Macintosh, Windows 3.1*x*, and Windows 95.

AT&T WorldNet Service can provide you with a connection to the Internet, complete with an e-mail address and Web browser software.

AT&T WorldNet Service is a pay-per-use Internet service provider. To sign on, you need a modem connected to your computer, a phone line, and a credit card to register.

If you currently use another Internet service provider, be aware that installing AT&T WorldNet Service software may change your computer's current Internet software configuration. You may not be able to access the Internet through your original service provider after you install AT&T WorldNet Service.

You need a registration number for completing the installation during setup. If you are an AT&T long-distance residential customer, please use this code: **L5SQIM631**. If you use another long-distance service, please use this registration code: **L5SQIM632**.

The CD includes versions of AT&T WorldNet Service software with either Netscape Navigator browser software or Microsoft Internet Explorer browser software. You can install either one, based on your preference.

Note: Windows 95 users can also find the latest version of Internet Explorer software in the Cool Stuff section, for use with the Microsoft NetMeeting software.

Installation: For Windows users, select the version of AT&T WorldNet Service software with your preferred Web browser from the CD interface product list and then click on the Install button to launch the installer program. Mac users should open the AT&T WorldNet Service folder within the Cool Stuff folder, open the folder for the version you want, and then double-click on the Install AT&T WorldNet icon to launch the installer.

White Pine Software's CU-SeeMe and Enhanced CU-SeeMe

In the CU-SeeMe category for Windows users; in the CU-SeeMe folder for Mac users. For Macintosh, Windows 3.1*x*, and Windows 95.

CU-SeeMe (for Windows 95) and Enhanced CU-SeeMe (for Windows 3.1*x* and Macintosh) is a leading Internet videoconferencing program that lets you connect to CU-SeeMe reflector sites and videoconference with as many as eight other people at one time (12 in the Windows 95 version). With CU-SeeMe and Enhanced CU-SeeMe, you can also share audio, text chat, and use a whiteboard to share pictures, drawings, and documents with other users. All three versions also let you connect directly to another person for private, one-on-one videoconferencing.

These demo versions of CU-SeeMe and Enhanced CU-SeeMe expire after 30 days of use. If you like CU-SeeMe and wish to find out more about purchasing a license, go to White Pine's CU-SeeMe Web page at `http://www.cu-seeme.com`.

Special Requirements: Although you can use CU-SeeMe as a *lurker* (someone who receives video but doesn't send video to others in a conference), you'll probably want to have a video camera like the Connectix QuickCam, to make full use of the program.

Installing: For Windows users, all you have to do to install CU-SeeMe or Enhanced CU-SeeMe is select the program from the product list and click on the Install button in the interface. This launches the program's installer, which leads you through the installation process. For Macintosh users, open the CU-SeeMe folder and double-click on the file named Install 1 Enhanced CU-SeeMe. This launches the installer and gets Enhanced CU-SeeMe set up on your machine.

For the Mac version only, you are prompted to enter a serial number while installing Enhanced CU-SeeMe. When you see this prompt, enter the following number: **DCBE00100Z6T2HCC**.

VocalTec's Internet Phone

In the Internet Phone category for Windows users; in the Internet Phone folder for Mac users. For Macintosh, Windows 3.1*x*, and Windows 95.

Internet Phone is a popular Internet telephony program, which uses an online directory of hundreds of Chat Rooms full of people with whom you can chat. You can also create your own, private Chat Rooms for your friends and associates to use. The Windows 95 version has a load of extra features, such as videoconferencing, text chat, file transfer, and a whiteboard for sharing files. All three versions of Internet Phone on the CD are demo versions that expire after a few weeks of use. If you like Internet Phone and want to purchase a license, you can get more information and order online at `http://www.vocaltec.com`.

Special Requirements: In the Internet Phone folder for Macintosh users, you'll see two folders with different versions of Internet Phone, labeled "Internet Phone" and "Internet Phone for 68K." The first of these is for users of Macintoshes or Mac clones with PowerPC CPUs; the other version is for Macintoshes with 68040 CPUs. Older Macs with 68030, 68020, or 68000 CPUs unfortunately can't use Internet Phone.

Installing: To install either Windows version of Internet Phone, just select the appropriate version from the product list and click on the Install button. This launches the installer and gets you set up to use Internet Phone. For Mac users, simply drag the appropriate folder from the Internet Phone folder on the CD onto your Mac's hard drive.

Netspeak WebPhone

In the WebPhone category list. For Windows 3.1*x* and Windows 95.

WebPhone 3.0 is an Internet telephony product with lots of features and a neat "flip phone" interface that resembles a cellular telephone. WebPhone has four "lines" so you can put calls on hold. It also supports multi-party conference calling, voice mail, and text chat. The Windows 95 version also includes a videoconferencing function.

The versions of WebPhone on the CD are limited functionality demos; they are limited to three minutes per call, and only one line can be used at a time. In addition, the video function of the Windows 95 version expires in 30 days. If you like WebPhone, you can find out more information and register the program online at `http://www.netspeak.com`.

Installing: In the CD interface, select the version of WebPhone appropriate for your operating system (Windows 3.1*x* or Windows 95) from the product list, and click on the Install button. This launches the installer, which guides you through the rest of the installation process.

Microsoft NetMeeting and Internet Explorer

In the NetMeeting category list. For Windows 95 only.

NetMeeting is a full-featured video, audio, and data conferencing application that is part of the Microsoft Internet Explorer product line. With NetMeeting, you can connect with another NetMeeting user (or a user of another H.323-compatible Internet telephony program such as Intel Internet Video Phone) and share audio and video. You can also use NetMeeting to connect to a group of other NetMeeting users in a data conference, and share information via text chat, file transfer, and whiteboard. Perhaps the neatest data conferencing feature is application sharing, which lets other participants in a conference edit and control documents in any Windows application running on one of the computers in the Meeting — pretty neat stuff.

As we mentioned, NetMeeting is part of the Internet Explorer family of software, and it works in conjunction with the Internet Explorer Web browser and e-mail program. If you don't already have Internet Explorer on your computer, you can find it in the Cool Stuff category.

NetMeeting has one of the greatest attributes of any software program — it's free — and the version you find on the CD is fully functioning, not a demo. You can find more information about NetMeeting on the Microsoft Web site, at `http://www.microsoft.com/netmeeting`.

Installing: Getting NetMeeting onto your computer is very easy. Just select Microsoft NetMeeting from the CD interface product list and click on the Install button. NetMeeting's Install Wizard launches and leads you through the rest of the installation.

To install Internet Explorer, click on the Cool Stuff category; then click on Internet Explorer in the product list and click on the Install button.

Adobe Acrobat Reader 3.0

In the Cool Stuff category for Windows users; in the Cool Stuff folder for Mac users. For Macintosh, Windows 3.1*x*, and Windows 95.

Adobe Acrobat reader is a utility program that lets you read documents which have been created in the portable document format (.pdf files). This file format can be read on various platforms without the need for special file converters. These PDF files let you see the documents with all of their graphics and text formatting, even though your computer doesn't have the program that created the documents.

We include Acrobat Reader on the CD because several of the programs on the CD use PDF files for product documentation and manuals. Even if you don't use these files on the CD, you may wish to install the Acrobat Reader anyway, because this file type is commonly found on the World Wide Web. In fact, the Acrobat Reader installer also includes a Web browser plug-in that lets you view PDF files directly in your browser window.

Installing: To install the Windows version of Adobe Acrobat Reader, simply select the version for your operating system (Windows 3.1 or Windows 95) from the Cool Stuff section and click on the Install button. Mac users should open the Adobe Acrobat Reader 3.0 folder within the Cool Stuff folder, and then double-click on the Install Acrobat Reader 3.0 file.

IDT Net2Phone

In the Cool Stuff category. For Windows 3.1*x* and Windows 95.

IDT Net2Phone is the client software for IDT's Net2Phone service, a gateway service that lets you make phone calls from your PC to regular telephones.

Special Requirements Unlike PC-to-PC Internet telephony software, Net2Phone is a for-pay service — you can't make calls to regular phones for free. You can, however, try out the software for free by making calls to toll-free 800 and 888 numbers. If you want to call other numbers, you need to purchase a calling plan from IDT — sort of like buying a prepaid calling card. For more information on registering Net2Phone and using it for calls to non-toll-free numbers, check out IDT's Web site at `http://www.net2phone.com`.

Installing: In the CD interface, select Net2Phone from the Cool Stuff list, and click on the Install button. This launches the installer, which guides you through the rest of the installation process.

If You've Got Problems (Of the CD Kind)

We tried our best to compile programs that work on most computers with the minimum system requirements. Alas, your computer may differ, and some programs may not work properly for some reason. The two likeliest problems are that you don't have enough memory (RAM) for the programs you want to use, or you have other programs running that affect the installation or running of the program. If you get error messages like `Not enough memory` or `Setup cannot continue`, try one or more of these methods and then try using the software again:

- Turn off any anti-virus software that you have on your computer. Installers sometimes mimic virus activity and may make your computer incorrectly believe that it is being infected by a virus.
- Close all running programs. The more programs you're running, the less memory is available to other programs. Installers also typically update files and programs. So if you keep other programs running, installation may not work properly.
- Make sure you have your Internet connection actively running. Some programs don't start properly if your Internet connection is inactive, or may not connect to their directory service systems correctly if the Internet connection is activated *after* the program has already been started.
- Have your local computer store add more RAM to your computer. If you're a Windows 95 or Power Macintosh user, adding more memory speeds things up and allows more programs to run at the same time.
- Turn on your operating system's virtual memory. Power Macintosh users actually *save* real RAM when virtual memory is on.
- If your Macintosh crashes with a system error (also called a *bomb*), or if Windows suffers a general protection fault (GPF), try restarting your Mac or exiting and rebooting Windows before using the software again.

If you still have trouble with installing the items from the CD, please call the IDG Books Worldwide Customer Service phone number: 800-762-2974 (outside the U.S.: 317-596-5261).

Index

• C •

• D •

• E •

• F •

• J •

• K •

• L •

(continued)

• Q •

• R •

• S •

• U •

• V •

• W •

• X •

• Y •

• Z •

AT&T WorldNet℠ Service

A World of Possibilities…

Thank you for selecting AT&T WorldNet Service — it's the Internet as only AT&T can bring it to you. With AT&T WorldNet Service, a world of infinite possibilities is now within your reach. Research virtually any subject. Stay abreast of current events. Participate in online newsgroups. Purchase merchandise from leading retailers. Send and receive electronic mail.

AT&T WorldNet Service is rapidly becoming the preferred way of accessing the Internet. It was recently awarded one of the most highly coveted awards in the computer industry, *PC Computing*'s 1996 MVP Award for Best Internet Service Provider. Now, more than ever, it's the best way to stay in touch with the people, ideas, and information that are important to you.

You need a computer with a mouse, a modem, a phone line, and the enclosed software. That's all. We've taken care of the rest.

If You Can Point and Click, You're There

With AT&T WorldNet Service, finding the information you want on the Internet is easier than you ever imagined it could be. You can surf the Net within minutes. And find almost anything you want to know — from the weather in Paris, Texas — to the cost of a ticket to Paris, France. You're just a point and click away. It's that easy.

AT&T WorldNet Service features specially customized industry-leading browsers integrated with advanced Internet directories and search engines. The result is an Internet service that sets a new standard for ease of use — virtually everywhere you want to go is a point and click away, making it a snap to navigate the Internet.

When you go online with AT&T WorldNet Service, you'll benefit from being connected to the Internet by the world leader in networking. We offer you fast access of up to 28.8 Kbps in more than 215 cities throughout the U.S. that will make going online as easy as picking up your phone.

Online Help and Advice
24 Hours a Day, 7 Days a Week

Before you begin exploring the Internet, you may want to take a moment to check two useful sources of information.

If you're new to the Internet, from the AT&T WorldNet Service home page at `www.worldnet.att.net`, click on the Net Tutorial hyperlink for a quick explanation of unfamiliar terms and useful advice about exploring the Internet.

Another useful source of information is the HELP icon. The area contains pertinent, time saving information-intensive reference tips, and topics such as Accounts & Billing, Trouble Reporting, Downloads & Upgrades, Security Tips, Network Hot Spots, Newsgroups, Special Announcements, etc.

Whether online or off-line, 24 hours a day, seven days a week, we will provide World Class technical expertise and fast, reliable responses to your questions. To reach AT&T WorldNet Customer Care, call **1-800-400-1447**.

Nothing is more important to us than making sure that your Internet experience is a truly enriching and satisfying one.

Safeguard Your Online Purchases

AT&T WorldNet Service is committed to making the Internet a safe and convenient way to transact business. By registering and continuing to charge your AT&T WorldNet Service to your AT&T Universal Card, you'll enjoy peace of mind whenever you shop the Internet. Should your account number be compromised on the Net, you won't be liable for any online transactions charged to your AT&T Universal Card by a person who is not an authorized user.*

*Today, cardmembers may be liable for the first $50 of charges made by a person who is not an authorized user, which will not be imposed under this program as long as the cardmember notifies AT&T Universal Card of the loss within 24 hours and otherwise complies with the Cardmember Agreement. Refer to Cardmember Agreement for definition of authorized user.

Minimum System Requirements

IBM-Compatible Personal Computer Users:

- IBM-compatible personal computer with 486SX or higher processor
- 8MB of RAM (or more for better performance)
- 15–36MB of available hard disk space to install software, depending on platform (14–21MB to use service after installation, depending on platform)
- Graphics system capable of displaying 256 colors
- 14,400 bps modem connected to an outside phone line and not a LAN or ISDN line
- Microsoft Windows 3.1*x* or Windows 95

Macintosh Users:

- Macintosh 68030 or higher (including 68LC0X0 models and all Power Macintosh models)
- System 7.5.3 Revision 2 or higher for PCI Power Macintosh models: System 7.1 or higher for all 680X0 and non-PCI Power Macintosh models
- Mac TCP 2.0.6 or Open Transport 1.1 or higher

- 8MB of RAM (minimum) with Virtual Memory turned on or RAM Doubler; 16MB recommended for Power Macintosh users
- 12MB of available hard disk space (15MB recommended)
- 14,400 bps modem connected to an outside phone line and not a LAN or ISDN line
- Color or 256 gray-scale monitor
- Apple Guide 1.2 or higher (if you want to view online help)
  If you are uncertain of the configuration of your Macintosh computer, consult your Macintosh User's guide or call Apple at 1-800-767-2775.

Installation Tips and Instructions

- If you have other Web browsers or online software, please consider uninstalling them according to the vendor's instructions.
- If you are installing AT&T WorldNet Service on a computer with Local Area Networking, please contact your LAN administrator for setup instructions.
- At the end of installation, you may be asked to restart your computer. Don't attempt the registration process until you have done so.

IBM-compatible PC users:
- Insert the CD-ROM into the CD-ROM drive on your computer.
- Select ***File/Run*** (for Windows 3.1*x*) or ***Start/Run*** (for Windows 95 if setup did not start automatically).
- Type ***D:\setup.exe*** (or change the "D" if your CD-ROM is another drive).
- Click ***OK***.
- Follow the onscreen instructions to install and register.

Macintosh users:
- Disable all extensions except Apple CD-ROM and Foreign Files Access extensions.
- Restart Computer.
- Insert the CD-ROM into the CD-ROM drive on your computer.
- Double-click the *Install AT&T WorldNet Service* icon.
- Follow the onscreen instructions to install. (Upon restarting your Macintosh, AT&T WorldNet Service Account Setup automatically starts.)
- Follow the onscreen instructions to register.

Registering with AT&T WorldNet Service

After you have connected with AT&T WorldNet online registration service, you will be presented with a series of screens that confirm billing information and prompt you for additional account set-up data.

The following is a list of registration tips and comments that will help you during the registration process.

I. Use one of the following registration codes, which can also be found in Appendix C of *Internet Telephony For Dummies,* 2nd Edition. Use L5SQIM631 if you are an AT&T long-distance residential customer or L5SQIM632 if you use another long-distance phone company.
II. During registration, you will need to supply your name, address, and valid credit card number, and choose an account information security word, e-mail name, and e-mail password. You will also be requested to select your preferred price plan at this time. (We advise that you use all lowercase letters when assigning an e-mail ID and security code, since they are easier to remember.)
III. If you make a mistake and exit or get disconnected during the registration process prematurely, simply click on "Create New Account." Do not click on "Edit Existing Account."
IV. When choosing your local access telephone number, you will be given several options. Please choose the one nearest to you. Please note that calling a number within your area does not guarantee that the call is free.

Connecting to AT&T WorldNet Service

When you have finished installing and registering with AT&T WorldNet Service, you are ready to access the Internet. Make sure your modem and phone line are available before attempting to connect to the service.

For Windows 95 users:
- Double-click on the ***Connect to AT&T WorldNet Service*** icon on your desktop.
  OR
- Select ***Start, Programs, AT&T WorldNet Software, Connect to AT&T WorldNet Service.***

For Windows 3.*x* users:
- Double-click on the ***Connect to AT&T WorldNet Service*** icon located in the AT&T WorldNet Service group.

For Macintosh users:
- Double-click on the ***AT&T WorldNet Service*** icon in the AT&T WorldNet Service folder.

Choose the Plan That's Right for You

The Internet is for everyone, whether at home or at work. In addition to making the time you spend online productive and fun, we're also committed to making it affordable. Choose one of two price plans: unlimited usage access or hourly usage access. The latest pricing information can be obtained during online registration. No matter which plan you use, we're confident that after you take advantage of everything AT&T WorldNet Service has to offer, you'll wonder how you got along without it.

Explore our AT&T WorldNet Service site at `http://www.att.com/worldnet`.

IDG Books Worldwide, Inc., End-User License Agreement

READ THIS. You should carefully read these terms and conditions before opening the software packet(s) included with this book ("Book"). This is a license agreement ("Agreement") between you and IDG Books Worldwide, Inc. ("IDGB"). By opening the accompanying software packet(s), you acknowledge that you have read and accept the following terms and conditions. If you do not agree and do not want to be bound by such terms and conditions, promptly return the Book and the unopened software packet(s) to the place you obtained them for a full refund.

1. **License Grant.** IDGB grants to you (either an individual or entity) a nonexclusive license to use one copy of the enclosed software program(s) (collectively, the "Software") solely for your own personal or business purposes on a single computer (whether a standard computer or a workstation component of a multiuser network). The Software is in use on a computer when it is loaded into temporary memory (RAM) or installed into permanent memory (hard disk, CD-ROM, or other storage device). IDGB reserves all rights not expressly granted herein.

2. **Ownership.** IDGB is the owner of all right, title, and interest, including copyright, in and to the compilation of the Software recorded on the disk(s) or CD-ROM ("Software Media"). Copyright to the individual programs recorded on the Software Media is owned by the author or other authorized copyright owner of each program. Ownership of the Software and all proprietary rights relating thereto remain with IDGB and its licensers.

3. **Restrictions on Use and Transfer.**

   (a) You may only (i) make one copy of the Software for backup or archival purposes, or (ii) transfer the Software to a single hard disk, provided that you keep the original for backup or archival purposes. You may not (i) rent or lease the Software, (ii) copy or reproduce the Software through a LAN or other network system or through any computer subscriber system or bulletin-board system, or (iii) modify, adapt, or create derivative works based on the Software.

   (b) You may not reverse engineer, decompile, or disassemble the Software. You may transfer the Software and user documentation on a permanent basis, provided that the transferee agrees to accept the terms and conditions of this Agreement and you retain no copies. If the Software is an update or has been updated, any transfer must include the most recent update and all prior versions.

4. **Restrictions on Use of Individual Programs.** You must follow the individual requirements and restrictions detailed for each individual program in Appendix C of this Book. These limitations are also contained in the individual license agreements recorded on the Software Media. These limitations may include a requirement that after using the program for a specified period of time, the user must pay a registration fee or discontinue use. By opening the Software packet(s), you will be agreeing to abide by the licenses and restrictions for these individual programs that are detailed in Appendix C and on the Software Media. None of the material on this Software Media or listed in this Book may ever be redistributed, in original or modified form, for commercial purposes.

5. Limited Warranty.

(a) IDGB warrants that the Software and Software Media are free from defects in materials and workmanship under normal use for a period of sixty (60) days from the date of purchase of this Book. If IDGB receives notification within the warranty period of defects in materials or workmanship, IDGB will replace the defective Software Media.

(b) IDGB AND THE AUTHORS OF THE BOOK DISCLAIM ALL OTHER WARRANTIES, EXPRESS OR IMPLIED, INCLUDING WITHOUT LIMITATION IMPLIED WARRANTIES OF MERCHANTABILITY AND FITNESS FOR A PARTICULAR PURPOSE, WITH RESPECT TO THE SOFTWARE, THE PROGRAMS, THE SOURCE CODE CONTAINED THEREIN, AND/OR THE TECHNIQUES DESCRIBED IN THIS BOOK. IDGB DOES NOT WARRANT THAT THE FUNCTIONS CONTAINED IN THE SOFTWARE WILL MEET YOUR REQUIREMENTS OR THAT THE OPERATION OF THE SOFTWARE WILL BE ERROR FREE.

(c) This limited warranty gives you specific legal rights, and you may have other rights that vary from jurisdiction to jurisdiction.

6. Remedies.

(a) IDGB's entire liability and your exclusive remedy for defects in materials and workmanship shall be limited to replacement of the Software Media, which may be returned to IDGB with a copy of your receipt at the following address: Software Media Fulfillment Department, Attn.: *Internet Telephony For Dummies, 2nd Edition*, IDG Books Worldwide, Inc., 7260 Shadeland Station, Ste. 100, Indianapolis, IN 46256, or call 800-762-2974. Please allow three to four weeks for delivery. This Limited Warranty is void if failure of the Software Media has resulted from accident, abuse, or misapplication. Any replacement Software Media will be warranted for the remainder of the original warranty period or thirty (30) days, whichever is longer.

(b) In no event shall IDGB or the authors be liable for any damages whatsoever (including without limitation damages for loss of business profits, business interruption, loss of business information, or any other pecuniary loss) arising from the use of or inability to use the Book or the Software, even if IDGB has been advised of the possibility of such damages.

(c) Because some jurisdictions do not allow the exclusion or limitation of liability for consequential or incidental damages, the above limitation or exclusion may not apply to you.

7. U.S. Government Restricted Rights. Use, duplication, or disclosure of the Software by the U.S. Government is subject to restrictions stated in paragraph (c)(1)(ii) of the Rights in Technical Data and Computer Software clause of DFARS 252.227-7013, and in subparagraphs (a) through (d) of the Commercial Computer@ndRestricted Rights clause at FAR 52.227-19, and in similar clauses in the NASA FAR supplement, when applicable.

8. General. This Agreement constitutes the entire understanding of the parties and revokes and supersedes all prior agreements, oral or written, between them and may not be modified or amended except in a writing signed by both parties hereto that specifically refers to this Agreement. This Agreement shall take precedence over any other documents that may be in conflict herewith. If any one or more provisions contained in this Agreement are held by any court or tribunal to be invalid, illegal, or otherwise unenforceable, each and every other provision shall remain in full force and effect.

Installation Instructions

To install the CD-ROM that comes with this book on your Windows computer:

1. **Insert the CD-ROM into your CD-ROM drive.**
2. **Windows 95 users: Click on the Start button.**

   Windows 3.*x* users: In the Program Manager, click on the File menu.
3. **Click on Run.**
4. **In the box that appears, type** `D:\SETUP.EXE`.

The first time you use the CD, you will see the License Agreement. After you agree to the terms of the agreement, the interface will open up and you can start browsing the CD.

To install the CD-ROM on your Macintosh computer:

1. **Insert the CD-ROM into your CD-ROM drive.**
2. **On the desktop, use your mouse to double-click on the "Internet Telephony For Dummies" CD icon.**

   A window of the same name opens on your screen.
3. **Double-click on the file "Read Me."**

   This file contains specific instructions for installing the included software.

Please see Appendix C for installation instructions for the programs on the CD.

IDG BOOKS WORLDWIDE REGISTRATION CARD

RETURN THIS REGISTRATION CARD FOR FREE CATALOG

Title of this book: Internet Telephony For Dummies®, 2E

My overall rating of this book: ❑ Very good [1] ❑ Good [2] ❑ Satisfactory [3] ❑ Fair [4] ❑ Poor [5]

How I first heard about this book:

❑ Found in bookstore; name: [6] ❑ Book review: [7]

❑ Advertisement: [8] ❑ Catalog: [9]

❑ Word of mouth; heard about book from friend, co-worker, etc.: [10] ❑ Other: [11]

What I liked most about this book:

What I would change, add, delete, etc., in future editions of this book:

Other comments:

Number of computer books I purchase in a year: ❑ 1 [12] ❑ 2-5 [13] ❑ 6-10 [14] ❑ More than 10 [15]

I would characterize my computer skills as: ❑ Beginner [16] ❑ Intermediate [17] ❑ Advanced [18] ❑ Professional [19]

I use ❑ DOS [20] ❑ Windows [21] ❑ OS/2 [22] ❑ Unix [23] ❑ Macintosh [24] ❑ Other: [25] ______ (please specify)

I would be interested in new books on the following subjects:
(please check all that apply, and use the spaces provided to identify specific software)

❑ Word processing: [26] ❑ Spreadsheets: [27]

❑ Data bases: [28] ❑ Desktop publishing: [29]

❑ File Utilities: [30] ❑ Money management: [31]

❑ Networking: [32] ❑ Programming languages: [33]

❑ Other: [34]

I use a PC at (please check all that apply): ❑ home [35] ❑ work [36] ❑ school [37] ❑ other: [38] ______

The disks I prefer to use are ❑ 5.25 [39] ❑ 3.5 [40] ❑ other: [41] ______

I have a CD ROM: ❑ yes [42] ❑ no [43]

I plan to buy or upgrade computer hardware this year: ❑ yes [44] ❑ no [45]

I plan to buy or upgrade computer software this year: ❑ yes [46] ❑ no [47]

Name: Business title: [48] Type of Business: [49]

Address (❑ home [50] ❑ work [51]/Company name:)

Street/Suite#

City [52]/State [53]/Zipcode [54]: Country [55]

❑ **I liked this book!** You may quote me by name in future IDG Books Worldwide promotional materials.

My daytime phone number is ______

YES!

Please keep me informed about IDG's World of Computer Knowledge. Send me the latest IDG Books catalog.

NO POSTAGE
NECESSARY
IF MAILED
IN THE
UNITED STATES

BUSINESS REPLY MAIL

FIRST CLASS MAIL PERMIT NO. 2605 FOSTER CITY, CALIFORNIA

IDG Books Worldwide
919 E Hillsdale Blvd, STE 400
Foster City, CA 94404-9691